THE NATION STATE OF ISRAEL

THE NATION STATE
OF ISRAEL

ROBERT H. SCHRAM

Copyright © 2021 by Robert H. Schram.

ISBN: Softcover   978-1-6641-7131-2
          eBook   978-1-6641-7130-5

All rights reserved. No part of this book may be reproduced or transmitted in any form or by any means, electronic or mechanical, including photocopying, recording, or by any information storage and retrieval system, without permission in writing from the copyright owner.

Any people depicted in stock imagery provided by Getty Images are models, and such images are being used for illustrative purposes only. Certain stock imagery © Getty Images.

Print information available on the last page.

Rev. date: 04/28/2021

To order additional copies of this book, contact:
Xlibris
844-714-8691
www.Xlibris.com
Orders@Xlibris.com
827296

CONTENTS

Introduction vii

Chapter 1 Geography and Environment 1
Chapter 2 The Economy and Culture 8
Chapter 3 Demographics 20
Chapter 4 Government and Legal
           System 28
Chapter 5 History from Abraham to
           1948 CE 42
Chapter 6 Zionism 56
Chapter 7 The Holocaust 63
Chapter 8 The Nation State of Israel 72

References 91

Introduction

Israel has archeological evidence of the earliest migration of hominids out of Africa and Canaanite tribes in the geography since the Middle Bronze Age (2,000–1,750 BCE). The Kingdoms of Israel and Judah emerged during the Near Eastern Iron Age I (1,200-1000 BCE). After emerging from Egyptian slavery c. 1300 BCE Joshua led the tribes across the Jordan River into the Promised Land. After about 300 years of chaotic rule by the Judges Israel appointed its first King Saul and began its United Monarchy over the kingdoms of Israel and Judah, during the reigns of Saul, David and Solomon (c. 1047 BCE to 930 BCE). On the succession of Solomon's son Rehoboam c. 930 BCE, the country split into two kingdoms: the Kingdom of Israel (including the cities of Shechem and Samaria) in the north and the Kingdom of Judah (with Jerusalem) in the south. The Neo-Assyrian Empire destroyed the Kingdom of Israel c. 722 BCE and the first diaspora began during which the ten northern tribes never returned. Judah was later conquered and exiled by the Babylonians under the rule of Nebuchadnezzar (c. 597 and 586 BCE) for their second diaspora. The Babylonians were conquered by the Persians and the Jews were allowed to return to their Promised Land by the Persian king Cyrus the Great (c. 538 BCE). Alexander the Great and his Hellenistic Empire conquered the land c. 332 BCE. The Maccabeans revolted and overthrew the Hellenistic Selucids c. 140 BCE and rededicated the Jerusalem Temple (celebrated by the Jewish Holiday of Chanukah). The Maccabean led Hasmoneans ruled until 63 BCE. Then the Romans conquered the land and ruled until c. 70 CE when they destroyed the Jerusalem Temple and exiled the Jews into their third and longest diaspora (1,878 years from 70 to 1948 CE).

The Roman Republic subsequently installed the Herodian dynasty in 37 BCE, and in 6 CE created the Roman province of Judea. Judea lasted as a Roman province until failed Jewish revolts resulted in widespread destruction, the expulsion of the Jewish population, and the renaming of the region from *Iudaea* to *Syria Palaestina*. Emperor Hadrian (117–138 CE) saw Judaism as the cause of continuous rebellions so he prohibited the Torah and the Hebrew calendar and executed Judaic scholars. The Torah was ceremonially burned on the Temple Mount. At the former Temple sanctuary he installed two statues, one of Jupiter, another of himself.

The Jewish–Roman wars were a series of large-scale Jewish revolts against the Roman Empire between 66 and 135 CE. The first was between 66-73 CE. The second was the Kitos War (115–117 CE) which was more of an ethno-religious one fought external to the Judea Province. The third was the Bar Kokhba revolt (132–136 CE). The first and the last were nationalist rebellions, striving to restore an independent Judean state. The complete victory of the Romans in all three wars dramatically reduced the major Jewish population into a dispersed and persecuted minority. Simon bar Kokhba, the commander of the third revolt, was acclaimed as a Messiah, a heroic figure who could restore Israel. The revolt established an independent state of Israel over parts of Judea for more than two years, but a Roman army made up of six full legions with auxilia and elements from up to six additional legions finally crushed it. The Romans then barred Jews from Jerusalem, except their attendance on Tisha B'Av. Although Jewish Christians hailed Jesus as the Messiah and did not support Bar Kokhba, they were barred from Jerusalem along with the rest of the Jews.

With the destruction of the Second Temple the face of Judaism changed. The major ruling group of priests, the Sadducees lost all their power and were rendered obsolete. The Pharisees were the rabbinic group whose power spread among the synagogues of different communities. Sacrificing animals transformed into worshiping God through prayer. Rabbinic Judaism spread throughout the Roman world and beyond its borders. Jewish culture flourished in Babylon and Galilee where work on the Talmud commenced. Rabbi Yohanan ben Zakkai obtained permission to open a Judaic school at Yavne; his students secretly smuggled him away from Jerusalem in a coffin and he authored the Mishnah, the core text of Rabbinical Judaism.

Jewish presence in the Promised Land region persisted over the centuries. In the 7th century CE, the Levant was taken by the Arabs from the Byzantine Empire and remained in Muslim control until the First Crusade of 1099, followed by the Ayyubid conquest of 1187. The Mamluk Sultanate of Egypt extended its control over the Levant in the 13th century until its defeat by the Ottoman Empire in 1517. During the 19th century, national awakening among Jews led to the establishment of the Zionist movement followed by immigration to Palestine.

In 1947, the United Nations adopted a Partition Plan for Palestine recommending the creation of independent Arab and Jewish states and an internationalized Jerusalem. The plan was accepted by the Jews and rejected by Arab leaders. In 1948 the Jewish Agency declared Israel as an independent State. Israel controlled most of the former Mandated territory after the 1948 Arab–Israeli War and the West Bank and Gaza were held by neighboring Arab states. Israel fought several wars with Arab countries, after 1948 and since the Six-Day War in June 1967 they held occupied territories including the West Bank, Golan Heights and the Gaza Strip. Efforts to resolve the Israeli–Palestinian conflict have not resulted in a final peace agreement, while Israel has signed peace treaties and has been recognized by other Arab countries.

Israel defines itself as a Jewish democratic state and the nation state of the Jewish people. Only Israel and Tunisia have liberal democracies in the Middle East and North Africa region with parliamentary systems, proportional representation, and universal suffrage. Israel has the highest standard of living in the Middle East, and ranks among the world's top countries by percentage of citizens with military training, percentage of citizens holding a tertiary education degree, research and development spending by GDP percentage, women's safety, life expectancy, innovativeness, and happiness.[2]

Chapter One
Geography and Environment

Israel declared its independence on May 14, 1948 and was admitted to the United Nations on May 11, 1949. The nation is located on the southeastern shore of the Mediterranean Sea and the northern shore of the Red Sea. It's land borders Lebanon to the north, Syria to the northeast, Jordan on the east and Egypt to the southwest. The West Bank and Gaza Strip are Palestinian territories to the east and west respectively. Israel, despite its relatively small size (20,770-22,072 sq. km or 8,019-8,522 sq. miles contains geographically diverse features and 2% water. The total area under Israeli control, including the military-controlled and partially Palestinian-governed territory of the West Bank, is 27,799 square kilometers (10,733 sq. miles).

The country's proclaimed capital and largest city is Jerusalem.

The economic and technological center is Tel Aviv.

The National Anthem is *Hatikvah* (The Hope), the currency is the new shekel, and the official language is Hebrew. Arabic is a recognized language. The government is a Unitary parliamentary republic with the Knesset as its Legislative branch. The 2008 population census was 7,412,200 and estimated to be 9,262,700 in 2020. The ethnic breakdown in 2019 was: 74.2% Jewish; 20.9% Arab; 4.8% Other. The religious breakdown in 2019 was: 74.2% Jewish; 17.8% Islam; 2.0% Christianity; 1.6% Druze; and 4.4% Other. The estimated Gross Domestic Product (GDP) is $372,314 billion (51st world ranking) with a per capital income of $40,336 (34th world ranking) in 2020. The Gini coefficient or Gini index (a measure of statistical dispersion intended to represent the income inequality or wealth inequality within a nation or any group of people with 0 as perfect equality and 1 as perfect inequality) in 2018 was 0.348 (the global Gini coefficient is estimated to be between 0.61 and 0.68). The Human Development Index (HDI) in 2019 was very high at 0.919 ranking 19th out of 189 countries and territories worldwide. The United Nations Development Program (UNDP) publishes the HDI ranking countries into four tiers of human development by combining measurements of life expectancy, education, and per-capita income in its annual Human Development Report.[1]

Israel is located in the Levant area of the Fertile Crescent region. Despite its small size, there are a variety of geographic features, from the Negev desert in the south to the inland fertile Jezreel Valley, mountain ranges of the Galilee, Carmel and toward the Golan in the north. The Israeli coastal plain on the shores of the Mediterranean is home to most of the nation's population. East of the central highlands lies the Jordan Rift Valley, which forms a small part of the 4,039 mile Great Rift Valley. Running along the Jordan Rift Valley from Mount Hermon through the Hulah Valley and the Sea of Galilee to the Dead Sea (the lowest point on earth's surface) is the Jordan River. The Arabah is further south ending with the Gulf of Eilat, part of the Red Sea. Makhteshim (makhtesh - singular) have steep walls of resistant rock surrounding a deep closed valley, which is usually drained by a single wadi) are unique to Israel and the Sinai Peninsula. The Ramon Crater in the Negev is the largest makhtesh in the world, which measures 25 by 5 miles.

The result of tectonic movements within the Dead Sea Transform (DSF) fault system created the Jordan Rift Valley. The DSF forms the transform boundary between the African Plate to the west and the Arabian Plate to the east. The Golan Heights and all of Jordan are part of the Arabian Plate, while the Galilee, West Bank, Coastal Plain, and Negev along with the Sinai Peninsula are on the African Plate. On average about every 400 years an earthquake along the Jordan Valley segment has occurred (31 BCE, 363 CE, 749 CE, and 1033 CE). The deficit in slip that has built up since the 1033 event is sufficient to cause an earthquake. Destructive earthquakes leading to serious loss of life strike about every 80 years. Israel has stringent construction regulations but many public buildings and about 50,000 residential buildings are older than the regulations and do not meet the new standards. They are expected to collapse in the next significant earthquake.

Israeli temperatures vary widely, especially during the winter months. Coastal areas (e.g., Tel Aviv and Haifa) have cool, rainy winters, and long hot summers...a typical Mediterranean climate. The Northern Negev and the Beersheba area have a semi-arid climate with hot summers, cool winters, and fewer rainy days than the Mediterranean climate. The Arava areas and the Southern Negev have a desert climate with very hot, dry summers, and mild winters with few days of rain. The highest temperature in the continent of Asia (54.0° C or 129.2° F) was recorded in 1942 at Tirat Zvi kibbutz in the northern Jordan River valley. Mountainous regions can be windy and cold. Jerusalem with an elevation of 750 meters or 2,460 feet will usually receive at least one snowfall each year. Rain is rare from May to September in the entire nation. Israel has developed water-saving technologies in water scarce areas such as drip irrigation. Israel is the leading nation per capita in solar energy with almost every household using solar panels for water heating. In the Mediterranean Basin Israel has the largest number of plant species per square meter of all the countries. Due to four different phytogeographic regions in the country, the flora and fauna of Israel are extremely diverse. There are 2,867 known species of plants in the country of which at least 253 species have been introduced and are non-native. There are 380 nature reserves.[3]

Chapter Two
The Economy and Culture

Economic data is available for the territory of Israel, including the Golan Heights, East Jerusalem, and Israeli settlements in the West Bank. The country is ranked 16th in the World Economic Forum's *Global Competitiveness Report* and 54th on the World Bank's *Ease of Doing Business* index. Israel also ranked 5th in the world by share of people in high-skilled employment. In Western Asia and the Middle East Israel is considered the most advanced in economic and industrial development. This very high technology boom and rapid economic development springs from a high quality university education and a highly motivated and educated populace.

The country has limited natural resources but due to the intensive development of the agricultural and industrial sectors since 1948

Israel is self-sufficient in food production, with the exception of grains and beef. In 2017 imports totaled $66.76 billion (including raw materials, military equipment, investment goods, rough diamonds, fuels, grain, and consumer goods). Exports totaled $60.6 billion including machinery and equipment, software, cut diamonds, agricultural products, chemicals, textiles and apparel) The Bank of Israel holds $113 billion of foreign-exchange reserves. The United States gives the country military aid and economic assistance in the form of loan guarantees, which account for about half of Israel's external debt (one of the lowest in the developed world). Israel is a lender in terms of net external debt (assets vs. liabilities abroad), which in 2015 stood at a surplus of $69 billion.

Israel's development of cutting-edge technologies in software, communications and the life sciences compare to America's Silicon Valley. Israel's expenditures on research and development as a percentage of GDP ranks 1st in the world and 5th in the 2019 Bloomberg Innovation Index. Israel ranks 1st in the world with 140 scientists, technicians, and engineers per 10,000 employees (the United States number is 85). Since 2004 there have been six Israeli Nobel Prize-winning scientists. The country has frequently been ranked as one of the countries with the highest ratios of scientific papers per capita in the world. It leads the world in stem-cell research papers per capita since 2000. Israeli universities are ranked among the top 50 world universities in computer science (Technion and Tel Aviv University), mathematics (Hebrew University of Jerusalem) and chemistry (Weizmann Institute of Science). Israel is second to the United States in having the largest number of startup companies in the world and the third-largest number of NASDAQ-listed companies after the United States and China. Research and Development centers have some of the biggest names in high-tech multi-national corporations (Intel, Microsoft, IBM, Google, Apple, Hewlett-Packard, Cisco Systems, Facebook, and Motorola). American investor Warren Buffett's holding company Berkshire Hathaway bought the Israeli company Iscar for $4 billion in 2007; its first acquisition external to the United States. In Futron's 2012 Space Competitiveness Index Israel was ranked ninth in the world. The Israel Space Agency coordinates all space research with scientific and commercial goals; they have designed and built at least 13 commercial, research, and spy satellites. Israel is ranked among the world's most advanced space systems. Shavit is a space launch vehicle making the country the eighth nation to have this capability since 1988.

Innovation in water conservation has been spurred on by the ongoing water shortage as drip irrigation was invented by Israelis along with substantial agricultural modernization. The country is at the technological forefront of desalination and water recycling. The Sorek desalination plant is the largest seawater reverse osmosis (SWRO) desalination facility in the world. By 2014, Israel's desalination programs provided roughly 35% of Israel's drinking water and it is expected to supply 70% by 2050. The country hosts an annual Water Technology and Environmental Control Exhibition & Conference (WATEC) that attracts thousands of people from across the world. As a result of innovations in reverse osmosis technology, Israel is set to become a net exporter of water in the coming years.

Israeli engineers are on the cutting edge of solar energy technology and its solar companies work on projects throughout the world. The country saves 8% of its electricity consumption per year because of solar energy used in heating. Ketura Sun is the country's first commercial solar field built in 2011 by the Arava Power Company on Kibbutz Ketura covering twenty acres. The field consists of 18,500 photovoltaic panels made by Suntech, which produces about 9 gigawatt-hours (GWh) of electricity per year. Over the next two decades the field will eliminate the production of some 125,000 metric tons of carbon dioxide. Arava Power Company has 8 projects to be built in the Negev valued at 780 million NIS (Israeli new shekel) or approximately $204 million. Israeli offshore gas fields began producing natural gas in 2004. The country also imported gas from Egypt via the al-Arish–Ashkelon pipeline that was terminated in 2012 due to the Egyptian Crisis of 2011–14. Two natural gas reserves were found in 2009 (Tamar) and in 2010 (Leviathan). The natural gas reserves in Tamar and Leviathan could make Israel energy secure for more than 50 years. Natural gas commercial production began in 2013 at the Tamar field.

Israel's trailblazing electric car company *Better Place* shut down in 2013. A modern electric car infrastructure involving a countrywide network of charging stations to facilitate the charging and exchange of car batteries was created in order to lower oil dependency and the fuel costs for hundreds of motorists that use cars powered only by electric batteries. The Israeli model was being studied by several countries with Denmark and Australia implementing the model. Israel has 19,224 kilometers or 11,945 miles of paved roads, and 3 million motor vehicles although numbers per 1,000 persons is relatively low

(365) for a developed country. Egged is the largest bus carrier of several operating many of the 5,715 buses on scheduled routes. Railways are government owned across 1,277 kilometers or 793 miles. The number of train passengers per year has grown exponentially and was 53 million in 2015. 7.5 million tons of cargo per year are also transported by rail.[4]

Ben Gurion Airport (near Tel Aviv) and Ramon Airport (near Eilat) handle all international air travel and several small domestic airports handle local air travel.

The Port of Haifa is the country's oldest and largest port while the Ashdod Port is one of the few deep-water ports in the world built on the open sea. The smaller Port of Eilat is situated on the Red Sea, and is used mainly for trading with Far East countries. Tourism, especially religious tourism, is an important industry enhanced by the country's temperate climate, beaches, archeological sites, historical sites, biblical sites, and unique geography. In 2017, a record of 3.6 million tourists visited Israel, yielding a 25 percent growth since 2016 and contributed greatly to the economy. Jews from different diaspora communities brought their cultural and religious traditions with them to Israel, creating a melting pot of Jewish customs and beliefs. Arab influences are present in many cultural spheres, such as architecture, music, and cuisine. Life revolves around the Hebrew calendar. Jewish holidays determine work and school holidays, and the official day of rest is Saturday, the Jewish Sabbath.

Poetry and prose in Hebrew is the primary source of Israeli literature; a small body of literature is also published in other languages such as English. All printed matter and published audio and video recordings must have two copies deposited in the National Library of Israel at the Hebrew University of Jerusalem. In 2016, 89 percent of the 7,300 books transferred to the library were in Hebrew. In 1966, Shmuel Yosef Agnon shared the Nobel Prize in Literature with German Jewish author Nelly Sachs. The 2017 *Freedom of the Press* annual report by Freedom House ranked Israel as the Middle East and North Africa's most free country, and 64th globally. The American Academy Awards has nominated ten Israeli films as finalists for Best Foreign Language Film since 1949. The country maintains a vibrant theater scene. The Habima Theatre in Tel Aviv was founded in 1918 and is Israel's oldest repertory theater company and national theater. The Israel Museum in Jerusalem is one of Israel's most important cultural institutions and houses the Dead Sea Scrolls, along with an extensive collection of Judaica and European art. Yad Vashem (the national Holocaust museum) is the world's central archive of Holocaust-related information. *Beit Hatfutsot* (The Diaspora House), at Tel Aviv University, is interactive and devoted to the history of worldwide Jewish communities. In addition to major large city museums there are high-quality art spaces in many towns and kibbutzim. Israel has the highest number of museums per capita in the world. Several museums are devoted to Islamic culture, including the Rockefeller Museum and the L. A. Mayer Institute for Islamic Art, both in Jerusalem. The Rockefeller Museum specializes in archaeological remains from the Ottoman Empire and other periods of Middle East history.

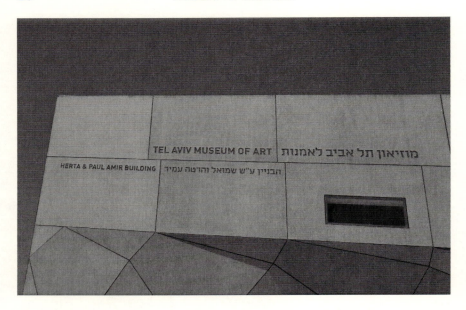

Music from all over the world has greatly influenced the country's music: Mizrahi, Sephardim, Hasidim, Greek, jazz, pop rock, etc. The Israel Philharmonic Orchestra is world-renowned and has been in operation since the founding of the state performing over two hundred concerts annually. Internationally acclaimed Israeli born musicians include: Itzhak Perlman, Pinchas Zukerman, and Ofra Haza. Israel has won the Eurovision Song Contest four times hosting it twice since 1973. The Red Sea International Jazz Festival has been hosted by the town of Eilat since 1987. The experiences of the pioneers who built the nation into a developed country are remembered in canonical folk songs; *Songs of the Land of Israel.*

About 50% of the Israeli-Jewish population attests to keeping kosher at home. Kosher restaurants were rare in the 1960s and now are about 25% of the total. Hotel eateries are likely to serve kosher food. The non-kosher retail market was once limited and grew significantly in the 1990s with the influx of post-Soviet immigration. Non-kosher fish, rabbits and ostriches are available and even pork (called *white meat*) is produced and consumed. Local cuisine and diaspora Jewish cuisine are popular since the establishment of the nation and since the 1970s an Israeli fusion cuisine has developed. Elements of the Mizrahi, Sephardi, and Ashkenazi cooking styles incorporates many foods traditionally eaten in the Levantine (Arab, Middle Eastern, and Mediterranean cuisines) such as falafel, hummus, shakshouka, couscous, and za'atar. Schnitzel, pizza, hamburgers, French fries are also common.

Association football and basketball are the most popular spectator sports. The premier football and basketball leagues are the Israeli Premier League and the Israeli Basketball Premier League. The largest football clubs are: Maccabi Haifa, Maccabi Tel Aviv, Hapoel Tel Aviv, and Beitar Jerusalem. Several clubs have competed in the Union of European Football

Associations (UEFA) Champions League and in 1970 the Israel football team qualified for the International Federation of Association Football (FIFA) World Cup. The 1974 Asian Games, held in Tehran, were the last Asian Games that had Israeli participation since many Arab countries refused to compete with Israel. Maccabi Tel Aviv B.C. has won the European championship in basketball six times. In 2016, the country was chosen as a host for the EuroBasket 2017. Chess is enjoyed by people of all ages with many Israeli grandmasters; Israeli chess players have won a number of youth world championships. Israel holds an annual international chess championship and hosted the World Team Chess Championship in 2005. Some schools include teaching chess since the Ministry of Education and the World Chess Federation have agreed to including it in school curriculums. Chess is taught in Beersheba kindergartens and the city has become a national chess center and is the home to the largest number of chess grandmasters of any city in the world. The Israeli chess team won the silver medal at the 2008 Chess Olympiad and the bronze medal among 148 teams at the 2010 Olympiad. Grandmaster Boris Gelfand won the Chess World Cup in 2009 and the 2011 Candidates Tournament for the right to challenge the world champion. Since its first Olympic medal win in 1992 the nation has won nine Olympic medals including the 2004 gold medal in windsurfing. The country has won over 100 gold medals in the Paralympic Games and is ranked 20th in the all-time medal count; it hosted the 1968 Summer Paralympics. The Maccabiah Games, an Olympic-style event for Israeli athletes, was inaugurated in the 1930s, and has been held every four years. A martial art (*Krav Maga*) developed in the ghettos during the Nazi Holocaust is used by the Israeli security forces and police. It is admired worldwide for its effectiveness and practical approach to self-defense.[5]

Chapter Three
Demographics

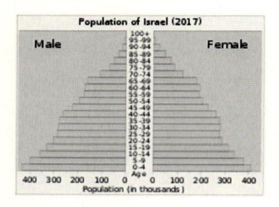

Since 2010 large numbers of migrant workers from Romania, Thailand, China, Africa, and South America have settled in Israel. Many are living illegally so their exact numbers are not known; their numbers are estimated from 166,000 to 203,000 people. About 92% of Israelis live in urban areas. The average life expectancy is 82.5 years (6th highest in the world).

Israel is often referred to as a Jewish state since it was established as a homeland for the Jewish people. The Law of Return grants citizenship to all Jews and those of Jewish ancestry. Retention of Israel's population since 1948 is about even in a country with mass immigration. Emigration from Israel (*yerida*) described as modest is primarily to the United States and Canada but is often cited by Israeli government ministries as a major threat to Israel's future.

Jews comprise three quarters of the population from diverse backgrounds. About 75% of Israeli Jews are *sabras,* i.e., *they* were born in Israel, while 16% are European and American immigrants, and 7% are immigrants from Asia and Africa (including the Arab world). European and Soviet Union Jews and their Israeli-born descendants, including the Ashkenazim, constitute about half of Jewish Israelis. Jews from Arab and Muslim countries and their descendants (including both Mizrahi and Sephardim), form most of the rest of the Jewish population. Jewish intermarriage rates run at over 35% with most from the Sephardim and Ashkenazim; over 25% of school children now originate from both communities.

The total number of Israeli settlers beyond the Green Line (pre-1967 border or the 1949

Armistice border) is over 600,000 (about 10% of the Jewish Israeli population). The name comes from the green ink used to draw the line on the map while the armistice talks were ongoing. After the 1967 Six-Day War, the territories captured by Israel beyond the Green Line came to be designated as East Jerusalem, the West Bank, Gaza Strip, Golan Heights, and Sinai Peninsula. The 1979 peace treaty returned the Sinai Peninsula to Egypt. These territories are often referred to as Israeli-occupied territories. In 2016, 399,300 Israelis lived in the West Bank settlements, and were re-established after the Six-Day War, in cities like Hebron and Gush Etzion bloc. There were more than 200,000 Jews living in East Jerusalem, and 22,000 in the Golan Heights. Approximately 7,800 Israelis lived in settlements in the Gaza Strip, known as Gush Katif, until they were evacuated by the government as part of its 2005 disengagement plan.

The four major metropolitan areas are: Gush Dan – the Tel Aviv metropolitan area (population 3,854,000), Jerusalem metropolitan area (population 1,253,900), Haifa metropolitan area (population 924,400), and Beersheba metropolitan area (population 377,100). Jerusalem is the largest municipality both in population and area. Tel Aviv and Haifa rank second and third as the most populous cities. There are 16 cities with populations over 100,000. There are 77 Israeli localities granted municipalities (or city) status by the Ministry of the Interior, four of which are in the West Bank. Two more cities are planned: Kasif, a planned city to be built in the Negev, and Harish, originally a small town that was built into a large city since 2015. Although the official language of Israel is Hebrew many other languages are spoken due to immigration. Arabic is common as is English, Russian, Amharic, and French. English was an official language prior to statehood and retains a role comparable to an official language. English is seen on road signs, official documents, and on many television programs. Many citizens communicate reasonably well in English, and it is taught from the early grades in elementary school. Israeli universities offer various courses in English.

The Holy Land of Israel is an important region to all Abrahamic faiths: Judaism, Christianity, Islam, Druze and Bahá'í. A 2016 Pew Research social survey indicates Israeli Jews vary widely: 49% self-identify as *Hiloni* (secular), 29% as *Masorti* (traditional), 13% as *Dati* (religious) and 9% as *Haredi* (ultra-Orthodox). Haredi Jews' growth rate is significant, and they are expected to represent more than 20% of Israel's Jewish population by 2028. Muslims make up about 17.6% of the population and are the country's largest religious minority. The 2% Christian population is composed primarily of Arab and Aramean Christians but also includes post-Soviet immigrants (about 300,000 of the one million Soviet immigrants are not Jewish), and followers of Messianic Judaism. Druze compose 1.6% of the population. Small numbers of many other religious groups include Buddhists and Hindus. Jerusalem is special to Jews, Muslims, and Christians with its religious sites: the Old City incorporates the Western Wall and the Temple Mount, the Al-Aqsa Mosque, and the Church of the Holy Sepulchre. Nazareth is holy in Christianity as the site of the Annuciation of Mary (the announcement of the Incarnation by the angel Gabriel to Mary). Tiberias and Safed are two of the Four Holy Cities in Judaism. The White Mosque in Ramla is holy to Muslims as the shrine of the prophet Saleh. The Church of Saint George in Lod is holy to both Christians and Muslims as the tomb of Saint George or Al Khidr). Other religious landmarks are in the West Bank: Joseph's Tomb in Nablus; the birthplace of Jesus and Rachel's Tomb in Bethlehem; the Cave of the Patriarchs in Hebron. The Bahá'í World Center in Haifa contains their administrative center of the Faith and the Shrine of the Báb (the leader of their faith buried in Acre).

South of the Baháʼí World Centre is Mahmood Mosque affiliated with the reformist Ahmadiyya movement. Mixed religious neighborhoods are not common but do exist in Haifa's Kababir (Jews and Ahmadi Arabs), with others in Jaffa, Acre, other Haifa neighborhoods, Harish, and Upper Nazareth.

From ancient times Judaism has always held education in high regard. Compulsory education was first introduced by Jewish communities in the Levant. The high quality of Israeli education has greatly enhanced economic development and its technological boom. In 2012, the country ranked third in the world in the number of academic degrees per capita (20 percent of the population). Israel has a school life expectancy of 16 years and a literacy rate of 97.8%. There are five types of schools established in 1953 by a State Education Law:

1. state secular,
2. state religious,
3. ultra-orthodox,
4. communal settlement schools,
5. Arab schools.

The largest group is the state secular attended by most Jewish and non-Arab pupils. Arabic is the language of instruction in the Arab schools where most Arabs attend. For children between the ages of three and eighteen education is compulsory. Schooling has three tiers – primary school (grades 1–6), middle school (grades 7–9), and high school (grades 10–12) that culminates with *Bagrut* matriculation exams. The *Bagrut* certificate requires proficiency in core subjects: mathematics, the Hebrew language, Hebrew and general literature, the English language, history, biblical scripture (replaced in Arab, Christian, and Druze schools by their own heritage), and civics. Forty-six percent of Israeli Jews hold post-secondary degrees and have an average of 11.6 years of schooling; they are one of the most highly educated of all major religious groups in the world. In 2014, 61.5% of all Israeli twelfth graders earned a matriculation certificate.

In higher education the country has nine public universities that are subsidized by the state and 49 private colleges. The Technion and the Hebrew University consistently ranked among world's 100 top universities by the prestigious *Academic Ranking of World Universities* (ARWU). Other major universities in the country include the Weizmann Institute of Science, Tel Aviv University, Ben-Gurion University of the Negev, Bar-Ilan University, the University of Haifa and the Open University of Israel. Ariel University, in the West Bank, is the newest university institution, upgraded from college status, and the first in over thirty years.[6]

Chapter Four
Government and Legal System

Israel has universal suffrage in its parliamentary democracy where a parliament member supported by a parliamentary majority becomes the prime minister...usually the chair of the largest party. The prime minister is the head of government and the cabinet. The Knesset is the 120-member parliament (four year terms) whose membership is based on proportional representation of the political parties which in practice has resulted in coalition governments with a 3.25% electoral threshold. West Bank Israeli settlers are eligible to vote and in the 2015 election 10 of the 120 (8%) Knesset members were settlers. When coalitions are unstable the Knesset can dissolve the government earlier than every four years by a no-confidence vote. The president of Israel is head of state, with limited and largely ceremonial duties. The

Basic Laws of Israel function as an uncodified constitution...various attempts to draft the formal document since 1948 (most recent in 2003) have fallen short of the mark, and instead Israel has evolved a system of basic laws and rights, which enjoy semi-constitutional status.

Israel defines itself as a *Jewish and democratic* state yet has no official religion, ergo creating a strong connection to Judaism and a conflict between state law and religious law. The balance between state and religion is kept similarly to what existed prior to statehood when Israel was under the British Mandate...i.e., interaction between the political parties keeps the balance. The Knesset on July 19, 2018 passed a Basic Law that characterizes the State of Israel as principally a *Nation State of the Jewish People* with Hebrew as its official language. The law gives *special status* to the Arabic language and gives Jews a unique right to national self-determination with the development of Jewish settlements *as a national interest*, empowering the government *to take steps to encourage, advance and implement this interest.*

Israel has a three-tier court system. The highest tier is the Jerusalem Supreme Court serving a dual role as the highest appeals court and the High Court of Justice that allows citizens and non-citizens to petition against state authorities' decisions. Just below the Supreme Court are district courts that are both courts of appeal and courts of first instance; i.e., people can petition against state authorities. The magistrate courts are the lowest on the three-tier situated in most cities across the country; they are situated in five of Israel's six districts. The legal system combines three legal traditions: English common law, civil law, and Jewish law based on the principle of *stare decisis* (precedent) and is an adversarial system, where the parties in the suit bring evidence before the court. Professional judges rather than juries, rule on court cases. Religious courts decide on issues of marriage and divorce: Jewish, Muslim, Druze, and Christian. Judges are elected by a committee of two Knesset members, three Supreme Court justices, two Israeli Bar members and two ministers (one of which, Israel's justice minister, is the committee's chairperson). The Knesset members of the committee are secretly elected by the Parliament, and traditionally one of them is an opposition member. The committee's Supreme Court justices are chosen by seniority. The Israeli Bar members are elected by the bar, and the second minister is appointed by the Israeli cabinet. The Administration of Courts (both General and Labor Courts) is situated in Jerusalem. The General and Labor Courts are paperless with electronic storage of court files and decisions. Israel's Basic Law: Human Dignity and Liberty seeks to defend human rights and liberties in Israel. As a result of *Enclave law*, large portions of Israeli civil law are applied to Israeli settlements and Israeli residents in the occupied territories.

The country is divided into six main administrative districts, known as *mehozot* (singular *mahoz*): Center, Haifa, Jerusalem, North, South, and Tel Aviv districts, as well as the Judea and Samaria Area in the West Bank. The Judea and Samaria Area and parts of the Jerusalem and Northern districts are not recognized internationally as part of Israel. Districts are further divided into fifteen sub-districts known as *nafot* (singular *nafa*), which are partitioned into fifty natural regions.

The administration and sovereignty in the former British territory of Mandatory Palestine (1923-1948) involves six territories and Israel proper: Gaza Strip, West Bank, East Jerusalem, West Jerusalem, Golan Heights. The Gaza Strip is administered by the Hamas led Palestinian National Authority (PA) (under Israeli occupation) whose recognition comes from witnesses to the Oslo II Accord. 137 UN members recognize the PA claim to sovereignty. The Area A in the West

Bank is led by Fatah led Palestinian National Authority (under Israeli occupation). West Bank Area B is administered by Fatah led PA and the Israeli military (under Israeli occupation). West Bank Area C is administered by the Israeli military (Palestinians under Israel occupation) and Israeli enclave law (Israeli settlements). East Jerusalem's governing authority (Israeli government) is recognized by Honduras, Guatemala, Nauru, and the United States. West Jerusalem's governing authority (Israeli government) is recognized by Australia, Russia, Czechia, Honduras, Guatemala, Nauru, and the United States. The United Nations claims its sovereignty as an international city along with East Jerusalem; a claim supported by various UN member states and the European Union and joint sovereignty is also widely supported. The Israeli government is the governing authority in the Golan Heights supported only by the United States. Syria claims Golan Heights sovereignty supported by all UN members except for the United States. Israel is governed by the Israeli government and 163 UN members approve of its governing authority and its sovereignty.

As a result of the Six-Day War in 1967 Israel captured and occupied the West Bank, East Jerusalem, the Gaza Strip, and the Golan Heights. The Sinai Peninsula was also captured and returned to Egypt as part of the 1979 Egypt–Israel Peace Treaty. For eighteen years (1982-2000) Israel occupied part of southern Lebanon (known as the Security Belt). Israeli settlements and military installations have been built within each territory except Lebanon. Under Israeli law the Golan Heights and East Jerusalem have been annexed and fully incorporated into the country...but not under international law. The territorial inhabitants have been granted permanent residency status and the ability to apply for citizenship. The UN Security Council has declared Israel's annexation of these two *occupied areas as null and void*. At times the status of East Jerusalem has been difficult in peace negotiations since Israel views it as part of its sovereign capital.

The West Bank excluding East Jerusalem is called Judea and Samaria in Israeli law and the nearly 400,000 Israeli settlers are considered part of Israel's population: they have Knesset representation; much of civil and criminal laws applies to them; their output is considered part of Israel's economy. Israel has refrained from annexing the West Bank without relinquishing its legal claim to the land or defining a border with the area. The perceived *demographic threat* of incorporating the West Bank Palestinian population into Israel is the rational for the political opposition to annexation. West Bank Palestinians cannot become Israeli citizens and it remains under Israeli military rule. Many in the international community consider this to be the longest military occupation in modern history. In 1950 Jordan occupied and annexed the West Bank following the Arab rejection of the UN decision to create two states in Palestine. Only Britain recognized Jordan's annexation and it has since ceded its claim to the territory to the Palestinian Liberation Organization (PLO). Israel administered the area with their military from 1967 until 1993. Since the Israel–PLO letters of recognition, most of population and cities have been under the internal jurisdiction of the Palestinian Authority, with partial Israeli military control. During periods of unrest Israel has redeployed its troops and reinstated full military administration. 87% of a West Bank barrier is inside the West Bank.

The United Nation's International Court of Justice, asserted its advisory opinion in 2004 on the legality of the construction of the West Bank barrier, and lands captured in the Six-Day War are occupied territories. UN Security Council Resolution 242, which emphasizes *the inadmissibility of the acquisition of territory by war* and calls on withdrawal from occupied territories so as to normalize relations with Arab states (known as the *Land for peace* principle). According to some, Israel has engaged in systematic and widespread human rights violations and war crimes against civilians. Some observers, such as Israeli officials, scholars, United States Ambassador to the UN Nikki Haley and UN secretary-generals Ban Ki-moon and Kofi Annan, assert that the UN is disproportionately concerned with Israeli misconduct.

Israel and Egypt operate a land, air, and sea blockade of the Gaza Strip and the international community considers Israel as an occupying power. Egypt first occupied the Gaza Strip from 1948 to 1967 and then by Israel after 1967. The unilateral disengagement plan in 2005 removed all Israeli settlers and forces from Gaza continuing to maintain control of its airspace and waters. Hamas assumed power over Gaza after the 2007 Battle of Gaza and control along the border was tightened with entering and exiting reserved for only humanitarian cases.

Israel has diplomatic relations with 161 UN member states and additionally with the Holy See, Kosovo, the Cook Islands and Niue. It has 107 diplomatic missions around the world; countries without diplomatic relations include most Muslim countries. There are six members of the Arab League that have normalized relations: Egypt and Jordan signed peace treaties in 1979 and 1994 respectively; Mauritania granted full diplomatic relations in 1999; Bahrain, the United Arab Emirates, Morocco, and Sudan granted full diplomatic relations in 2020. Lebanon, Syria, Saudi Arabia, Iraq, Iran, and Yemen are enemy countries under Israeli law, and Israeli citizens may not visit without permission from the Ministry of the Interior. China maintains good ties with both Israel and the Arab world. The United States and the Soviet Union were the first two countries to recognize the State of Israel almost simultaneously. The Soviet Union's diplomatic relations were broken in 1967 (post Six-Day War) and renewed in October 1991. Israel is considered the *most reliable partner in the Middle East*, by the United States and since 1967 the U.S. has provided over $100 billion in military assistance and grants. Due to the Mandate for Palestine the United Kingdom's relationship is seen as *neutral*. By 2007, Germany had paid 25 billion euros in reparations to the Israeli state and individual Israeli Holocaust survivors. Israel is included in the European Union's European Neighborhood Policy (ENP), which aims at bringing the EU and its neighbors closer. Turkey recognized the country in 1949 and established full diplomatic relations in 1991. Relations between Greece and Israel have improved since 1995 due to the decline of Israeli–Turkish relations (after the Gaza War [2008-2009] and the Gaza flotilla raid [5-31-2010]). Greece has a defense cooperation agreement with Israel and in 2010 the Israeli Air Force hosted a joint exercise. Greece has strong ties with Cyprus and the joint Cyprus-Israel oil and gas explorations centered on the Leviathan gas field enhanced Israel's relationship. Cyprus ties have been strengthened by its cooperation with the world's longest subsea electric power cable.

Azerbaijan which supplies Israel with oil is one of the few majority Muslim countries to develop bilateral strategic and economic relations. Israel has helped modernize the armed forces of Azerbaijan. Full diplomatic relations with India were established in 1992 and has fostered a strong military, technological, and cultural partnership. India is considered the most pro-Israel country in the world according to a 2009 opinion survey on behalf of the Israel Ministry of Foreign Affairs. India is the largest customer of the Israeli military equipment and Israel is the second-largest military partner of India after Russia. Due to common political, religious, and security interests, Ethiopia is Israel's main ally in Africa. Israel provides expertise to Ethiopia on irrigation projects and thousands of Ethiopian Jews live in Israel.

Israel has a history of providing emergency aid and humanitarian response teams to disasters across the world. In 1955 a foreign aid program began in Burma and the focus subsequently shifted to Africa. In 1957 the *Mashav* was established for International Development Cooperation (IDF). The program developed good will in the African continent but after the 1967 war relations weakened and the foreign aid program shifted to Latin America. Since the late 1970s Israel's foreign aid has gradually decreased and its African aid was reestablished. Additional humanitarian and emergency response groups work with the government: IsraAid, a joint program run by 14 Israeli organizations and North American Jewish groups; ZAKA; The Fast Israeli Rescue and Search Team (FIRST); Israeli Flying Aid (IFA); Save a Child's Heart (SACH); and Latet focusing on poverty and food insecurity. The IDF search and rescue unit sent 24 delegations to 22 countries between 1985 and 2015. The UN has set a target of 0.7% of gross national income (GNI) for foreign aid (in 2015 only six nations reached this target). Israel ranked 38th in the 2018 World Giving Index at 0.07% of GNI.[7]

The sole military wing of the Israeli security forces is the Israel Defense Forces (IDF) consisting of the army, air force, and navy headed by its Chief of General Staff, the *Ramatkal*, subordinate to the Cabinet. The current structure consolidated paramilitary organizations (chiefly the *Haganah*) during the 1948 Arab–Israeli War. *Aman* is the Military Intelligence Directorate that provides resources to the IDF which works with *Mossad* (overseas intelligence) and *Shabak* (internal intelligence). The IDF has been involved in several major wars and border conflicts in its relatively short history, making it one of the most battle-trained armed forces in the world. Service in the military is mandatory and most Israeli citizens are drafted into the military at the age of 18; men serve two years and eight months and women two years. After full time service men join the reserve forces with up to several weeks of reserve duty every year until forty-plus years of age. Most women are exempt from reserve duty. Arab citizens of Israel (except the Druze) and those engaged in full-time religious studies are exempt from military service, a long-time source of contention. *Sherut Leumi* (national service) is an alternative for those with military exemptions which involves a program of service in hospitals, schools, and other social welfare programs. The country has one of the world's highest percentage of citizens with military training resulting from its conscription program. The IDF has about 176,500 active troops and about 465,000 reservists.

High-tech weapons systems designed and manufactured in Israel and some foreign imports are key to Israel's military successes. The world has a few operational anti-ballistic missile systems and the Israeli Arrow missile is one of them. The Python air-to-air missile series has been most effective and considered crucial. The Spike missile is one of the most widely exported anti-tank guided missiles (ATGMs) in the world. The Iron Dome anti-missile air defense system gained worldwide acclaim after intercepting hundreds of Qassam, 122 mm Grad and Fajr-5 artillery rockets fired by Palestinian militants from the Gaza Strip. The *Ofeq* program has successfully developed a network of reconnaissance satellites. Israel is one of seven countries capable of launching satellites. Israel has not signed the Treaty on the Non-Proliferation of Nuclear Weapons and maintains a policy of deliberate ambiguity toward its nuclear capabilities. It is widely believed to possess nuclear, chemical, and biological weapons of mass destruction. The Navy's Dolphin submarines are believed to be armed with nuclear Popeye Turbo missiles, offering second-strike capability. Since the 1991 Gulf War when Iraqi Scud missiles rained down on civilians, all homes are required to have a *Merkhav Mugan* (reinforced security room impermeable to chemical and biological substances).

Military expenditure has always constituted a significant portion of the country's gross domestic product, with a peak of 30.3% of GDP spent on defense in 1975. Israel ranked 6th in the world by defense spending as a percentage of GDP, with 5.7% in 2016 and 15th by total military expenditure, with $18 billion. Israel ranked 5th globally for arms exports in 2017 although the majority of its arms export are unreported for security reasons. In the Global Peace Index Israel is consistently rated low, ranking 144th out of 163 nations in 2017.[8]

Chapter Five

History from Abraham to 1948 CE

The whole Israeli region was known as Palestine under the British Mandate (1920–1948) and with independence in 1948, the country formally took Ben-Gurion's suggested name 'State of Israel' after rejecting other proposed historical and religious names including *Eretz Israel* ('the Land of Israel'), *Ever* (from ancestor Eber), *Zion, and Judea*. Israel is often interpreted as 'struggle with God' after Hosea 12:4. The names 'Land of Israel' and 'Children of Israel' have historically been used to refer to the biblical Kingdom of Israel and the entire Jewish people. The patriarch Jacob was given the name Israel after he successfully struggled (wrestled) with the angel of God and his twelve sons became the twelve tribes of Israel. Jacob and his family were forced to go to Egypt due to famine in Canaan staying for 430 years. Jacob's descendants were enslaved for their last 200 years in Egypt when Jacob's great-great grandson Moses led them on an 'Exodus' out of slavery back to the 'Promised Land' in Canaan. The earliest known archaeological artifact to mention the word 'Israel' as a collective is the Merneptah Stele of ancient Egypt (dated to the late 13th century BCE).

During the Late Bronze Age (1550–1200 BCE), large parts of Canaan were vassal states that payed tribute to the Egyptian Kingdom with its administration in Gaza. The current archeological account of the Israelite culture posits that the region was not overtaken by force, but instead branched out with the Canaanite peoples and their cultures through the development of a distinct monolatristic (belief in many gods, but consistently worship only one deity)—and later monotheistic—religion centered on *Yahweh*. The archaeological evidence indicates a society of village-like centers with limited resources, small populations (300-400), and self-sufficiency through farming and herding. Economic interchange was common, and writing was used and recorded. No archeological evidence yet if there was ever a United Monarchy but there is evidence of the Canaanites in the Middle Bronze Age (2100–1550 BCE). Historians and archaeologists agree that a Kingdom of Israel existed by c. 900 BCE and that a Kingdom of Judah existed by c. 700 BCE.

Abraham, originally Abram (dob c. 1750 BCE) became the first Patriarch (in Judaism, Christianity, and Islam) after God told him to leave his father's (Terah's) house and settle in the land originally given to Canaan but then is promised to Abraham's progeny. A famous *Midrash* states that his father owned an idol store, and one-night Abraham destroyed all the idols except one, the largest, into whose hands Abraham placed a large stick. The next morning his father was incensed at the destruction of his merchandise and asked Abraham if he knew what happened? Abraham told him that the largest idol took the stick in his hands and destroyed all the other smaller idols. His father replied: *That is ridiculous, these idols are not alive...they are made of wood!* Abraham replied: *That is the whole point father!* Abraham's first son Ishmael with maidservant Hagar is promised that he will be the founder of a great nation (many Mideastern Arabs) and his second son Isaac, born of half-sister Sarah will inherit the 'Promised Land' given to Abraham by God. Abraham purchases a tomb at Hebron (the Cave of the Patriarchs) for Sarah's grave, thus establishing his right to the land. His heir Isaac marries Rebekah, a woman of his own kin, thus ruling the Canaanites out of any inheritance. Abraham later marries Keturah and has six more sons. He is buried beside Sarah and Isaac receives *all Abraham's goods*, while the other sons receive only *gifts* (Genesis 25:5-8).

The story continues with Isaac and Rebekah birthing twins first born Esau and Jacob. Jacob, later named Israel after struggling with an angel tricks his father Isaac who is old and blind into giving him his blessing that belonged to Esau as the firstborn son to become the leader of the family. Jacob had twelve sons and one daughter by four women, his wives, Leah and Rachel, and his concubines, Bilhah and Zilpah, who were, in order of their birth: Reuben, Simeon, Levi, Judah, Dan, Naphtali, Gad, Asher, Issachar, Zebulun, Joseph, and Benjamin. Daughter Dinah's ranking in the birth order is not known. The twelve sons became the heads of their own family groups, later known as the Twelve Tribes of Israel. Jacob displayed favoritism among his wives and children; he preferred Rachel and her sons, Joseph, and Benjamin, causing tension within the family that culminated in the sale of Joseph by his brothers into Egyptian slavery. Joseph through his dream interpretation became a trusted confidante of Egypt's Pharaoh. During a severe drought and famine in Canaan, Jacob and his descendants eventually moved to Egypt with the help of his son Joseph. The descendants of the Twelve Tribes were in Egypt for over 400 years and enslaved for over 200 years until Moses led them out of slavery to their 'Promised Land' receiving the Torah on Mount Sinai in c. 1300 BCE. (Gen. 25-50)

Moses died on Mount Nebo in Moab, on the eastern side of the River Jordan and Joshua led the Twelve Tribes into Canaan and took over the land through force in the Book of Joshua. The Kingdom was first ruled by Judges c. 1382-1063 BCE: Othniel, Ehud, Shamgar, Deborah, Gideon, Abimelech, Tola, Jair, Jephthah, Ibzan, Elon, Abdon, Samson. The Book of Judges' story is cyclical as the Israelites practice idolatry and God's punishment is oppression by foreigners. As the people entreat God for help, He sends judges to deliver them. After a period of peace, the cycle is repeated. Other themes in the Book are: that God does not always do what is expected of him; foreigners consistently underestimate God and Israel; the judges are flawed and not adequate for their task; the progressive disunity of the Israelite community. The Book leaves out the Ark of the Covenant, tribal cooperation, and no mention of a central shrine and the High Priest. The Book also has passages and themes that represent anti-monarchist views. The judges come from the Spirit of God and not from prominent dynasties, nor through elections or appointments. Anti-monarchist theology is most apparent toward the end of the Gideon cycle as he refuses to create a dynastic monarchy as the Israelites beg him to do; the rest of Gideon's lifetime saw peace in the land. When his son Abimelech ruled Shechem as a power-craved tyrant he was guilty of much bloodshed (Chapters 8-9). The stories of Samson, Micah, and Gibeah highlight the violence and anarchy of decentralized rule. The rare exception to the rule that *great biblical women were married or related to great men* is the prophetess and judge Deborah who stands exclusively on her own merits...her husband's name was Lapidot. Deuteronomy contains the laws by which Israel is to live in the Promised Land. The Book of Joshua chronicles the conquest of Canaan and its allotment among the tribes. The Book of Judges describes the settlement of the land. The Book of Samuel describes the consolidation of the land and people under King David. The Book of Kings describes the destruction of kingship, and loss of the land. The final tragedy in the Book of Kings results from Israel's failure to uphold its part of the covenant, i.e., faithfulness to God brings economic, military, and political success, but unfaithfulness brings defeat and oppression.

The two Books of Kings provides a history of Israel and Judah from the death of King David to the release of Jehoiachin from Babylon imprisonment (c. 960 – c. 560 BCE). The Jerusalem Bible divides the two books of Kings into eight sections:

- 1 Kings 1:1–2:46 = The Davidic Succession
- 1 Kings 3:1–11:43 = Solomon in all his glory
- 1 Kings 12:1–13:34 = The political and religious schism
- 1 Kings 14:1–16:34 = The two kingdoms until Elijah
- 1 Kings 17:1 – 2 Kings 1:18 = The Elijah cycle
- 2 Kings 2:1–13:25 = The Elisha cycle
- 2 Kings 14:1–17:41 = The two kingdoms to the fall of Samaria
- 2 Kings 18:1–25:30 = The last years of the kingdom of Judah

David's son King Solomon's reign marks a key event in Israeli history, the construction of the First Temple in Jerusalem 480 years after the Exodus from Egypt. As a consequence of Idolatry the kingdom of David is split in two in the reign of his own son Rehoboam, who becomes the first to reign over the kingdom of Judah in the south. In the north, Israeli dynasties come and go rapidly, and idolatry continues so God brings the Assyrians to destroy the northern kingdom, leaving Judah as the sole custodian of the Promise. The 14th king of Judah, Hezekiah institutes far reaching reforms, centralizes Jerusalem Temple sacrifices, and destroys idolatrous images. God prevents an Assyrian invasion. Because of the apostasy of the next king Manasseh who reverses Hezekiah's reforms God announces he will destroy Jerusalem. Manasseh's righteous grandson Josiah reinstitutes Hezekiah's reforms, but it is too late. Through the prophetess Huldah God affirms that Jerusalem is to be destroyed after the death of Josiah. In the final chapters the Neo-Babylonian Empire of King Nebuchadnezzar razes Jerusalem, destroys the Temple, and the people are led into Babylonian exile[9].

The Kingdom of Israel was destroyed c. 721 BCE, when it was conquered by the Neo-Assyrian Empire (first exile and diaspora). Babylonian King Nebuchadnezzar II in 586 BCE conquered Judah, destroyed Solomon's Temple and exiled the Jews to Babylon (second exile and diaspora). This defeat of Judah is collaborated in the Babylonian Chronicles. After Medo-Persian Cyrus the Great conquered Babylon he returned the exiled Israelites to Judah beginning in 538 BCE. The Second Temple was constructed c. 520 BCE. The former Kingdom of Judah became the province of Judah (*Yehud Medinata*) under Persian rule with borders covering a smaller territory. Archeological surveys show a reduced population of about 30,000 people in the 5th to 4th centuries BCE.

The autonomous Judah province gradually developed back into an urban society dominated by Judeans. In c. 333 - 331 BCE Alexander the Great invaded the Middle East (conquering and ruling Israel and Palestine). During Greek occupation there was extensive growth and development that included urban planning and the establishment of well-built fortified cities. Incorporated into the Ptolemaic and finally the Seleucid empires, the southern Levant was heavily Hellenized (pottery, trade, and commerce), building tensions between Judeans and Greeks. The most Hellenized areas were Ashkelon, Jaffa, Jerusalem, Gaza, and ancient Nablus. The conflict erupted in 167 BCE with the Maccabean Revolt, resulting in an independent Hasmonean Kingdom in Judah (celebrated as Chanukah), expanding to much of modern Israel, as the Seleucids gradually lost control in the region.

The Roman Republic invaded the region in c. 63 BCE and took control of Syria, and then intervened in the Hasmonean Civil War. The struggle between pro-Roman and pro-Parthian (also Arsacid) Judean factions eventually led to the installation of Herod the Great and consolidation of the Herodian kingdom as a vassal Judean state of Rome. Judea became a Roman province with violent struggles between the Israelis and Romans, culminating in the Jewish–Roman wars. The wars ended in wide-scale destruction, expulsions, genocide, and enslavement of masses of Jewish captives. An estimated 1,356,460 Jews were killed as a result of the First Jewish Revolt (66-70 CE). The Second Jewish Revolt (115–117 CE) led to the death of more than 200,000 Jews. The Third Bar Kokhba Jewish Revolt (132–136 CE) resulted in the death of 580,000 Jewish soldiers. After Bar Kokhba's defeat the Jewish population was drastically reduced as the third and longest exile and diaspora was under way. There was a continuous small Jewish presence and Galilee became its religious center. The Mishnah and part of the Talmud were composed during the 2nd to 4th centuries CE in Tiberias, Jerusalem, and Babylon. The region became populated predominantly by Greco-Romans on the coast and Samaritans in the north[10].

Christianity was gradually evolving over Roman Paganism, when the area stood under Byzantine rule (continued Roman rule in the east with Constantinople as its capital). Through the 5th and 6th centuries, the dramatic events of the repeated Samaritan revolts reshaped the land, with massive destruction to Byzantine Christian and Samaritan societies and a resulting decrease of the population. After the Persian conquest and the installation of a short-lived Jewish Commonwealth in 614 CE, the Byzantine Empire reconquered the country in 628 CE. The Arabs who adopted Islam conquered the region, including Jerusalem between 634–641 CE. Over the next three centuries control of the region transferred between the Rashidun Caliphs, Umayyads, Abbasids, Fatimids, Seljuks, Crusaders, and Ayyubids.

The First Crusade (1099 CE) sieged Jerusalem massacring 60,000 people including 6,000 Jews who sought refuge in a synagogue. There were Jewish communities all over the country including Jerusalem, Tiberias, Ramleh, Ashkelon, Caesarea, and Gaza. Jews fought alongside Saracen (Fatimid) troops and Arabs until forced to retreat by the Crusader fleet and land army. In the twelfth century the Spanish-Jewish poet Yehuda Halevi migrated to the Land of Israel and called on all Jews to do the same (1141 CE). World renowned Rabbi Maimonides (Rambam) visited Jerusalem and prayed on the Temple Mount (1165 CE). Sultan Saladin, founder of the Ayyubid dynasty, defeated the Crusaders in the Battle of Hattin and then captured

Jerusalem and most of Palestine (1187 CE). Saladin issued a proclamation that invited Jews to return and settle in Jerusalem which they did. Some historians compare Saladin's decree inviting Jews to return with the one issued by the Persian king Cyrus the Great over 1,600 years earlier when he defeated the Babylonians ending the second exile and diaspora. Over 300 rabbis from France and England including Rabbi Samson ben Abraham of Sens migrated to Israel in 1211 CE strengthening the Jewish communities. The 13th century Spanish Rabbi Nachmanides (Ramban), considered the leader of world Jewry, greatly praised the Land of Israel and viewed Jewish settlement in the Land as a positive commandment incumbent on all Jews writing:

> *If the gentiles wish to make peace, we shall make peace and leave them on clear terms; but as for the land, we shall not leave it in their hands, nor in the hands of any nation, not in any generation.*

The Mamluk sultans of Egypt took over in 1260 CE ruling the area between Cairo and Damascus and initiated some development along the postal road connecting the two cities. Jerusalem, without the protection of city walls since 1219 CE experienced new construction projects centered around the Al-Aqsa Mosque compound on the Temple Mount. The Mamluk

Sultan Baybars converted the Hebron Cave of the Patriarchs in 1266 CE into an exclusive Islamic sanctuary and banned Christians and Jews from entering. The ban remained in place until Israel took control of the building in 1967. In the closing years of the 15th century, Safed and its environs had developed into the largest concentration of Jews in Palestine. With the help of the Sephardic immigration from Spain, the Jewish population had increased to 10,000 by the early 16th century. From 1516 until 1918 the region was ruled by the Ottoman Empire. Britain defeated the Turkish forces and set up a military administration across the former Ottoman Syria. Safed and Tiberias were destroyed in 1660 during a Druze revolt. Local Arab Sheikh Zahir al-Umar created a de facto independent Emirate in the Galilee in the late 18th century. The Ottoman Turks regained control of the area after Sheikh Zahir al-Umar died. Napoleon's troops were repelled from their assault on Acre in 1799 that prompted the French to abandon the Syrian campaign. Egyptian conscription and taxation policies caused Palestinian Arab peasants to revolt in 1834 led by Muhammad Ali. The revolt was suppressed, and Ottoman rule was restored with British support in 1840. Shortly after, the Tanzimat reforms were implemented throughout the Ottoman Empire. After the Allies conquered the Levant during

World War I, the territory was divided between Britain and France under the mandate system in 1920. The British-administered area included modern day Israel and was named Mandatory Palestine[11].

Chapter Six
Zionism

The Assyrian diaspora in 722 BCE resulted in the scattering of the ten northern Israeli tribes all over the Middle East; they were never found or returned. In the second Babylonian diaspora in 597 and 586 BCE the Jews remained a unified community in Babylon and others settled in the Nile Delta of Egypt, and other parts of the Middle East. When Babylon was conquered by the Persians (Cyrus the Great) the Jews were allowed to return to their homeland (c. 538 BCE). The Jews who had settled in Egypt became mercenaries in Upper Egypt on Elephantine Island and retained their religion, identity, and social customs under their own laws during the both the Persian and Greek occupations.

Judea became a protectorate of the Roman Empire in 63 CE under the administration of a governor and a king; the governor's business was to regulate trade and maximize tax revenue. The governors would maximize their revenue so they could pocket for themselves as much as possible. Governorships were bought at high prices enhancing their desire to exploit the populace for their own gain. Even with a Jewish king, the Judeans revolted in 70 CE and the Roman response was overwhelming, destroying Jerusalem, the Second Temple, and systematically driving the Jews out of the country for their third and longest diaspora (70-1948 CE). On the top of an isolated rock plateau (a mesa) overlooking the Dead Sea, Herod the Great built two palaces and *Masada* (fortress) for himself between 37-31 CE. At the end of the First Jewish–Roman War 960 *Sicarii* rebels (men, women, and children) retreated to *Masada;* the Romans besieged the fort for two years (73-74 CE). As the fortress was about to be breached the rebels all committed suicide rather than surrender to the Romans.

After 74 CE the Jews were spread out in Africa, Asia, and Europe. During this 1,878-year-diaspora many Jews aspired to return to Zion and the Land of Israel. Some communities settled in Palestine after they were expelled from Spain in 1492 CE. During the 16th century, Jewish communities settled in Jerusalem, Tiberias, Hebron, and Safed. Rabbi Yehuda Hachasid led a group of 1,500 to Jerusalem in 1697. In the second half of the 18th century, Eastern European opponents of Hasidism, known as the *Perushim,* settled in Palestine.

The First *Aliyah* (migration to Israel) to Ottoman-ruled Palestine, began in 1881, as Jews fled pogroms in Eastern Europe. The First Aliyah was the cornerstone for widespread Palestinian Jewish settlements. Many settlements were established from 1881 to 1903 purchasing about 350,000 dunams (about 86,487 acres) of land. In the diaspora Hebrew was used primarily for religious purposes and a means to communicate whenever Jews had different native tongues. Russian born Eliezer Ben-Yehuda migrated to Jerusalem in 1881 and successfully encouraged the revival of the Hebrew language which became the national tongue. A Hebrew school system emerged, and new words were coined or borrowed from other languages for modern inventions and concepts.

The Austro-Hungarian journalist Theodor Herzl is known as the father of modern political Zionism and the State of Israel. Political Zionism offered a solution to the so-called European Jewish question in conformity with the goals and achievements of other national projects of the time. Herzl wrote *Der Judenstaat* (The Jewish State) in 1896 that accurately predicted what occurred fifty-two years later:

> *Therefore, I believe that a wonderous generation of Jews will spring into existence. The Maccabeans will rise again. Let me repeat once more my opening words: The Jews wish to have a State, and they shall have one. We shall live at last as*

free men on our own soil and die peacefully in our own home. The world will be freed by our liberty, enriched by our wealth, magnified by our greatness. And whatever we attempt there to accomplish for our own welfare will react with beneficent force for the good of humanity.

After the Kishinev pogrom (killing, raping, and destroying Jewish homes by the ruling Russians what is now Moldova) a Second *Aliyah* (1904–14) settled 40,000 Jews in Palestine. Both the First and Second *Aliyah* migrants were mainly Orthodox Jews, although the Second *Aliyah* included socialist groups who established the *Kibbutz* movement (communal settlements). The *Kibbutzim* were mostly communal agricultural settlements, and the first Hebrew city Tel Aviv was established in 1909. The settlements defended themselves with armed self-defense organizations; the first such organization was *Bar-Giora,* a small secret guard founded in 1907. In 1909 the larger *Hashomer* organization was founded and replaced *Bar-Giora*. British Foreign Secretary Arthur Balfour sent his Balfour Declaration to Baron Rothschild (Walter Rothschild, 2nd Baron Rothschild) during the first World War. The Declaration stated that Britain intended to create a *Jewish national home* in Palestine.

The Jewish Legion (primarily Zionist volunteers) assisted with the British conquest of Palestine in 1918. The Arabs opposed British rule and Jewish immigration leading to the 1920 Palestine riots and the formation of a Jewish militia, the *Haganah* (The Defense) in 1920 as an outgrowth of *Hashomer.* The Paramilitary groups, *Irgun* and *Lehi,* or the Stern Gang eventually split off. The League of Nations in 1922 granted Britain the Mandate for Palestine that included the Balfour Declaration with similar provisions regarding the Arab Palestinians. In 1922 Arabs and Muslims dominated the area's population with Jews accounting for about 11%, and Arab Christians about 9.5%.

The Third (1919–23) and Fourth *Aliyahs* (1924–29) brought 100,000 more Jews to Palestine. The rise of Nazism and its increasing Jewish persecution led to the Fifth *Aliyah* in the 1930s with 250,000 Jewish immigrants. The Fifth *Aliyah* accompanied by land purchases was a major cause of the 1936-39 Arab revolt: several hundred Jews and British security personnel were killed. The British authorities alongside the *Haganah* and *Irgun* killed 5,032 Arabs and wounded 14,760. Over ten percent of the adult male Palestinian Arab population were killed, wounded, imprisoned, or exiled. The British 1939 White Paper introduced restrictions on

Jewish immigration. Jewish refugees fleeing the Holocaust were turned away by countries around the world and a clandestine movement (*Aliyah Bet*) was organized to bring Jews to Palestine. By the end of World War II, the Jewish population of Palestine had increased to 33% of the total population.[12]

Chapter Seven
The Holocaust

The Holocaust (aka the Shoah) was the genocide of European Jews during World War II (1941-1945) throughout German-occupied countries. Nazi Germany with collaborators systematically murdered about two-thirds of Europe's Jews and one-third of the world's Jews (6 million men, women, and children). There were about 9.5 million Jews in Europe in 1933: 3.5 million in Poland; 3 million in the Soviet Union; nearly 800,000 in Romania; 700,000 in Hungary; over 500,000 in Germany. The murders occurred in pogroms, mass shootings, extermination camps, gas chambers, and gas vans. The persecution and extermination evolved starting with concentration camps for political opponents and 'undesirables' in 1933 (Dachau was the first on 3-22-1933). In 1933

Hitler was given plenary powers by the 'Enabling Act' and the government began isolating Jews from society. At first, they boycotted Jewish businesses. In 1935 the Nuremberg Laws outlawed mixed marriages with Jews, took away Jewish citizenship, and prevented Jews free participation in the economy. In April 1933, the Law for the Restoration of the Professional Civil Service was passed excluding Jews and other non-Aryans from the civil service. Jews were disbarred from practicing law, being editors or proprietors of newspapers, joining the Journalists' Association, or owning farms. In 1933 there were about 50,000 Jewish-owned German businesses, and about 7,000 were still Jewish-owned in April 1939. Works by Jewish composers, authors, and artists were excluded from publications, performances, and exhibitions. Jewish doctors were dismissed or urged to resign. *Kristallnacht* (Night of Broken Glass) was two nights of ransacking Jewish businesses throughout Germany and Austria (November 9–10, 1938). Over 7,500 Jewish shops (out of 9,000) were looted and attacked, and over 1,000 synagogues damaged or destroyed. The Nazis set up ghettos to segregate Jews in 1939 with thousands of camps and detention sites across German-occupied Europe. The final stage of the genocide was called the *Final Solution to the*

Jewish Question, discussed by senior Nazis at the Wannsee Berlin Conference (January 1942). All anti-Jewish measures were radicalized and coordinated by the Gestapo (SS). Killings were committed throughout occupied Europe and in territories of Germany's allies. *Einsatzgruppen* (Paramilitary death squads) with the German Army and local collaborators, murdered about 1.3 million Jews during World War II in pogroms and mass shootings. Victims were forced into sealed freight trains and deported from European ghettos to extermination camps. The Nazis and their collaborators murdered others totaling about 11 million ethnic Poles, Soviet civilians, prisoners of war, the Roma, the handicapped, political-religious dissidents, and gay men.

In the 'Nazi Genocidal State' bureaucrats identified Jews, confiscated property, and scheduled deportations. Companies terminated Jewish employment and later used them as slaves. Jewish faculty and students were dismissed from universities. Pharmaceutical companies tested drugs on camp prisoners and other companies built crematorias. All personal property was recycled or reused by the state. Valuables stolen from victims were laundered through a concealed account of the German

National Bank. Some Christian churches defended converted Jews, but otherwise, Saul Friedländer wrote in 2007:

> Not one social group, not one religious community, not one scholarly institution or professional association in Germany and throughout Europe declared its solidarity with the Jews ...

Medical experiments on at least 7,000 inmates were conducted and twenty-three senior physicians and others were charged at the Nuremberg Tribunals (13 trials from 1945-1949) for crimes against humanity (including the head of the German Red Cross, tenured professors, clinic directors, and biomedical researchers). Experiments took place at Auschwitz, Buchenwald, Dachau, Natzweiler-Struthof, Neuengamme, Ravensbrück, Sachsenhausen, and elsewhere dealing with: male and female sterilization; war wound treatment; counteracting chemical weapons; new vaccines and drugs; survival under harsh conditions. The *euthanasia decree* later called *Aktion T4*, became law in 1939 and the number of mentally and physically handicapped persons murdered was about 150,000. There were six camps used exclusively for extermination:

Camp	Location (occupied Poland)	Deaths
Auschwitz II	Brzezinka	1,082,000 (all Auschwitz camps; includes 960,000 Jews)
Bełżec	Bełżec	600,000
Chełmno	Chełmno nad Nerem	320,000
Majdanek	Lublin	78,000
Sobibór	Sobibór	250,000
Treblinka	Treblinka	870,000
Total		**3,200,000**

Other camps sometimes described as extermination camps include Maly Trostinets near Minsk (occupied Soviet Union), where about 65,000 died by shooting and in gas vans; Mauthausen in Austria; Stutthof, near Gdańsk, Poland; and Sachsenhausen and Ravensbrück in Germany. According to the *Encyclopedia of Camps and Ghettos*, there were 23 main Nazi concentration camps and most had a system of satellite camps that brought the total to at least 1,000. The main 23 camps were:

1. Arbeitsdorf concentration camp,
2. Auschwitz concentration camp,

3. Bergen-Belsen concentration camp,
4. Buchenwald concentration camp,
5. Dachau concentration camp,
6. Flossenbürg concentration camp,
7. Gross-Rosen concentration camp,
8. Herzogenbusch concentration camp,
9. Hinzert concentration camp,
10. Kaiserwald concentration camp,
11. Kauen concentration camp,
12. Kraków-Płaszów concentration camp,
13. Majdanek concentration camp,
14. Mauthausen concentration camp,
15. Mittelbau-Dora concentration camp,
16. Natzweiler-Struthof concentration camp,
17. Neuengamme concentration camp,
18. Niederhagen concentration camp,
19. Ravensbrück concentration camp,
20. Sachsenhausen concentration camp,
21. Stutthof concentration camp,
22. Vaivara concentration camp,
23. Warsaw concentration camp

By 1942 several resistance groups were formed: the Jewish Combat Organization (ŻOB) and Jewish Military Union (ŻZW) in the Warsaw Ghetto, and the United Partisan Organization in Vilna. Over 100 revolts and uprisings occurred in at least 19 ghettos and elsewhere in Eastern Europe. The best known was the Warsaw Ghetto Uprising in April 1943. There were revolts in other ghettos and camps where Nazis were killed: Treblinka, the Bialystok Ghetto, Sobibor, the *Sonderkommando* at Auschwitz.

It is estimated that 20,000-100,000 Jews participated in partisan units throughout Europe and the occupied Polish and Soviet territories.

Country of Origin	Death toll of Jews
Albania	591
Austria	65,459
Baltic states	272,000
Belgium	28,518
Bulgaria	11,393
Croatia	32,000
Czechoslovakia	143,000
Denmark	116
France	76,134
Germany	165,000
Greece	59,195
Hungary	502,000
Italy	6,513
Luxembourg	1,200
Netherlands	102,000
Norway	758
Poland	2,100,000
Romania	220,000

Country of Origin	Death toll of Jews
Serbia	10,700
Soviet Union	2,100,000
Total	**5,896,577**

Group	Estimate of people killed during Holocaust era (1933–1945)
Soviet civilians (excl. 1.3 million Jews)	5.7 million
Soviet POWs (incl. c. 50,000 Jewish soldiers)	3 million
Non-Jewish Poles	c. 1.8 million
Serb civilians	312,000
Handicapped	Up to 250,000
Roma	196,000–220,000
Jehovah's Witnesses	c. 1,900
Criminals and "asocials"	at least 70,000
Gay men	Hundreds; unknown
Political opponents, resistance	Unknown

The government of Israel requested reparations of $1.5 billion from the Federal Republic of Germany in 1951 to finance the rehabilitation of 500,000 Jewish survivors; after negotiations $845 million was paid. West Germany allocated another $125 million in 1988. Germany set up the *Remembrance, Responsibility and Future* Foundation in 2000 for a number of companies (BMW, Deutsche Bank, Ford, Opel, Siemens, and Volkswagen) that used forced labor (slaves) during the war and paid out €4.45 billion to former slave laborers (up to €7,670 each). In 2013 Germany provided €772 million to fund nursing care, social services, and medication for 56,000 Holocaust survivors around the world. The French state-owned railway company, the SNCF, agreed in 2014 to pay $60 million to Jewish-American survivors (about $100,000 each), for its role in the transport of 76,000 Jews from France to extermination camps between 1942 and 1944.[13]

Chapter Eight
The Nation State of Israel

When the Second World War ended in 1945 countries were unwilling to accept the surviving European Jews. At every Passover Seder since 70 CE a return to the Jewish Promised Land was noted by the statement: *Next year in Jerusalem*. The United Kingdom with its Mandate over Palestine implemented Jewish immigration limits that neither Jews (limits too low) or Palestinian Arabs (limits too high) liked. Jews began a guerrilla campaign as the *Haganah* joined *Irgun and Lehi* in an armed struggle against British rule. The *Haganah* attempted to bring the hundreds of thousands of Jewish Holocaust survivors and refugees to Palestine by ship in the program *Aliyah Bet.* The Royal Navy intercepted most of the ships detaining the refugees in Atlit and Cyprus.

The British conducted a series of widespread raids in 'Operation Agatha' and in response *Irgun* attacked the Jerusalem King David Hotel that housed Britain's administrative headquarters (July 22, 1946) killing 91 and injuring 46 people of various nationalities. The insurgency continued throughout the rest of 1946 and 1947 despite Britain's repressive efforts to stop it or by bringing the recalcitrant Jews and Arabs to the negotiating table. The Palestine issue was brought to the newly formed United Nations in February 1947. The Report of the UN General Assembly Special Committee on Palestine (September 3, 1947) proposed a plan to replace the British Mandate with *an independent Arab State, an independent Jewish State, and the City of Jerusalem under an International Trusteeship System*. The insurgency peaked in July 1947, with a series of widespread guerrilla raids culminating in the 'sergeants affair'. Britain had sentenced three *Irgun* fighters to death for the Acre Prison break in May that freed 27 *Irgun and Lehi* militants and captured two British sergeants threatening their deaths if the three *Irgun* fighters were executed. After the three were executed, the *Irgun* killed the two sergeants and hung their bodies from eucalyptus trees, booby-trapping one of them with a mine which injured a British officer as he cut the body down. The hangings caused widespread British outrage and were a major factor in the consensus forming that it was time to evacuate Palestine.

The British cabinet (September 1947) decided to evacuate Palestine since the Palestine Mandate was no longer tenable. Colonial Secretary Arthur Creech Jones believed that four major factors led to the evacuation decision:

1. Jewish and Arab negotiators were unwilling to compromise regarding the question of a Jewish state in Palestine;
2. The economic pressure of continuing to station a large garrison in Palestine to deal with an expanding Jewish insurgency and Arab rebellion;
3. The "deadly blow to British patience and pride" caused by the hangings of the sergeants;
4. The mounting government criticism for its failure to find a new Palestine policy in place of the White Paper of 1939.

UN General Assembly Resolution 181 (II) (November 29, 1947) recommended the adoption and implementation of the *Plan of Partition with Economic Union* (essentially all that was proposed by the Special Committee on Palestine on September 3rd). The Jewish Agency, representing the Jewish community, accepted the plan and the Arab League and Arab Higher Committee of Palestine rejected it. The Arab Higher Committee proclaimed a three-day strike, and riots broke out in Jerusalem (December 1, 1947) that grew into a civil war. Arab militias and gangs attacked Jewish areas defended mainly by the *Haganah, Irgun and Lehi.* Through April 1948 250,000 Palestinian Arabs fled or were expelled, due to a number of factors. The day before the expiration of the British Palestine Mandate May, 14 1948) David Ben-Gurion, the head of the Jewish Agency, declared:

> *The establishment of a Jewish state in Eretz-Israel, to be known as the State of Israel.*

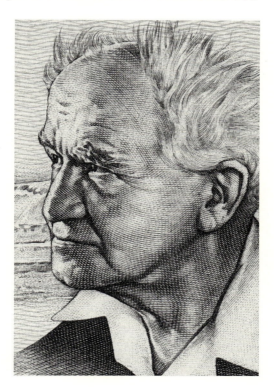

The '1948 Arab–Israeli War' began the next day on May 15, 1948. The armies of four Arab countries: Egypt, Syria, Transjordan, and Iraq and contingents from Yemen, Morocco, Saudi Arabia, and Sudan entered what had been British Mandatory Palestine. Apparently, the invading purpose was to prevent the establishment of a Jewish state, but some Arab leaders talked about driving the Jews into the sea and many Jews believed their real intent was to slaughter all Jews. A ceasefire was declared one year later with temporary borders, known as the Green Line. Jordan annexed what became known as the West Bank, including East Jerusalem, and Egypt occupied the Gaza Strip. The UN estimated that more than 700,000 Palestinians were expelled by or fled from advancing Israeli forces during the conflict—what would become known in Arabic as the *Nakba* (catastrophe). About 156,000 remained and became Arab citizens of Israel.[14]

By a United Nations majority vote, Israel became a member of the UN on May 11, 1949. The kibbutzim communities were pivotal in establishing the new state and Prime Minister David Ben-Gurion dominated politics with his Labor Zionist movement. Israel and Jordan were interested in a peace agreement, but

Britain did not want to damage British interests in Egypt, so it ended any progress.

The non-government sponsored *Mossad LeAliyah Bet* (Institute for Immigration B) organized illegal and clandestine immigration during the late 1940s and early 1950s before it was disbanded in 1953. There was both open and clandestine assistance for immigrants especially in the Middle East and Eastern Europe where Jewish lives were endangered and leaving was difficult. The immigrants came for different reasons: Zionist beliefs, promise of a better life, to escape persecution, and expulsion. Holocaust survivors and Jews from Arab and Muslim countries increased the Jewish population to two million by 1958. Those immigrants without possessions were housed in temporary tent cities known as *ma'abarot* with over 200,000 people by 1952. European Jews often were treated more favorably than others and generally spent less time in the transit camps. During this austerity period, food, clothes, and furniture had to be rationed. The crisis led Prime Minister Ben-Gurion to sign a reparations agreement with West Germany; mass protests ensued against accepting monetary compensation for the Holocaust.

Israeli civilians during the 1950s were frequently attacked by Palestinian *fedayeen*, mostly from the Egyptian-occupied Gaza Strip. The attacks led to several Israeli counterraids. The Egyptians nationalized the Suez Canal and blockaded Israeli shipping from the Canal and the Straits of Tiran in 1956. The blockade, increasing *fedayeen* attacks, and Arab grave threats prompted Israel to declare war by attacking Egypt. A secret alliance between Israel, Britain, and France overran the Sinai Peninsula and withdrew under UN pressure for a guarantee that Israel would have complete shipping rights in the Red Sea via the Straits of Tiran and the Canal. The 'Suez Crisis' significantly reduced Israeli border infiltration. A Nazi war criminal Adolf Eichmann was captured in Argentina in 1960, brought to trial in Israel, and executed in 1962.

Starting in 1964 Arab countries tried to divert the headwaters of the Jordan River to deprive Israel of water resources as Israel planned to divert Jordan River water into the coastal plain provoking tensions. Egyptian President Gamal Abdel Nasser refused to recognize Israel and called for its destruction leading the cause with other Arab nationalists. Israeli-Arab relations had deteriorated by 1966 to the point of actual battles occurring between Israeli and Arab forces. The UN peacekeepers, stationed in the Sinai Peninsula since 1957 were expelled by Egypt in May 1967 blocking Israel's access to the Red Sea. At the same time Egypt massed its army near Israel's border. Other Arab states mobilized their forces and Israel called the actions a *casus belli* and on June 5th launched a pre-emptive strike against Egypt. Jordan, Syria, and Iraq responded and attacked Israel. In a 'Six-Day War': Israel defeated Jordan and captured the West Bank; defeated Egypt and captured the Gaza Strip and Sinai Peninsula; defeated Syria and captured the Golan Heights. Jerusalem's boundaries grew and incorporated East Jerusalem, and the 1949 Green Line became the administrative boundary between Israel and the occupied territories.

Following the 1967 war there was a 1967–1970 'War of Attrition' as the Arab League adopted the *Three No's* resolution and Israel faced attacks from the Egyptians in the Sinai Peninsula, and from Palestinian groups targeting Israelis in the occupied territories, and around the world. The Palestinian Liberation Organization (PLO), established in 1964 was the most important Palestinian group; it committed itself to *armed struggle as the only way to liberate the homeland*. Palestinian groups launched a wave of attacks (late 1960s and early 1970s) against Israeli and

Jewish targets worldwide, including, including a massacre of Israeli athletes at the 1972 Munich Summer Olympics. Israel responded with an assassination campaign against the massacre organizers and bombed the PLO headquarters in Lebanon.

The 'Yom Kippur War' started on October 6, 1973 as the Egyptian and Syrian armies launched a surprise attack in the Sinai Peninsula and Golan Heights. The war ended on October 25 as Israel successfully repelled the Egyptian and Syrians. Israel lost over 2,500 soldiers and the war collectively took 10–35,000 lives. The government was exonerated of responsibility for failures before and during the war, but public anger forced Prime Minister Golda Meir to resign. In July 1976, an airliner was hijacked during its flight from Israel to France by Palestinian guerrillas and landed at Entebbe, Uganda. Israeli commandos carried out an operation in which 102 out of 106 Israeli hostages were successfully rescued.

Menachem Begin's Likud party took control from the Labor Party in the 1977 Knesset elections. Later in 1977 Egyptian President Anwar El Sadat became the first Arab head of state to recognize Israel as he spoke before the Israeli Knesset. Sadat and Begin signed the 'Camp David Accords' (1978) and the 'Israel–Egypt Peace Treaty' (1979). Israel withdrew from the Sinai Peninsula and agreed to enter autonomy negotiations for West Bank and Gaza Strip Palestinians. A PLO guerilla raid from Lebanon on March 11, 1978 led to the 'Coastal Road massacre' and in response Israel invaded southern Lebanon to destroy the PLO bases south of the Litani River. Israel secured southern Lebanon until a UN force and the Lebanese army took over. The PLO soon resumed its policy of attacks against Israel. The PLO infiltrated the south over the next several years and kept up a sporadic shelling across the border. In retaliation, Israel attacked by air and ground. Prime Minister Begin provided incentives for Israelis to settle in the occupied West Bank, increasing friction in the area. International controversy over the status of Jerusalem was inflamed when the government passed the 'Basic Law' in 1980 that Jerusalem is the Capital of Israel and reaffirmed to some Israel's 1967 annexation of Jerusalem by government decree. The territory of Israel has not been defined by Israeli legislation and no act specifically included East Jerusalem therein. In numerous resolutions, most UN member states declared that actions taken by Israel to settle its citizens in the West Bank, and impose its laws and administration on East Jerusalem, are illegal and have no validity. Internationally the annexation of the Golan Heights in 1981 was not recognized. Israel's population diversity and size expanded in the 1980s and 1990s with several waves of Ethiopian Jews and immigration from the post-Soviet states.

The Israeli air force destroyed Iraq's sole nuclear reactor (June 7, 1981) being constructed outside of Baghdad, to impede Iraq's nuclear weapons program. In 1982 the PLO launched a series of attacks and missiles into northern Israel and Israel invaded Lebanon to destroy their bases. In six days of fighting, the PLO Lebanese military forces were destroyed, and the Syrians were decisively defeated. The 'Kahan Commission' (Israeli government inquiry) later held Begin and several generals indirectly responsible for the Sabra and Shatila massacre and held Defense minister Ariel Sharon as bearing *personal responsibility* (he resigned). Israel bombed PLO headquarters in Tunisia in 1985 in response to a Palestinian terrorist attack in Cyprus. In 1986 Israel withdrew from most of Lebanon maintaining a borderland buffer zone in southern Lebanon until 2000 to engage with Hezbollah. The 'First Intifada' was in 1987 against Israeli rule by the Palestinians with violence and waves of uncoordinated demonstrations in the occupied West Bank and Gaza. From

1987 to 1993 the Intifada was more organized and included economic and cultural measures aimed at disrupting the Israeli occupation with more than one-thousand people killed. During the 1991 Gulf War Israel heeded American calls to not participate as the PLO supported Iraqi Saddam Hussein and his Scud missile attacks against Israel.

Yitzhak Rabin became Prime Minister in 1992 on his party's platform calling for compromise with Israel's neighbors. Israeli Shimon Peres and PLO Mahmoud Abbas signed the 'Oslo Accords' in 1993 giving the Palestinian National Authority the right to govern parts of the West Bank and the Gaza Strip as the PLO recognized Israel's right to exist and pledged an end to terrorism. The '1994 Israel–Jordan peace treaty' made Jordan the second Arab country to normalize relations. Israel's continued settlements, checkpoints, and economic deterioration damaged Arab public support for the Accords. Israeli public support for the Accords was also damaged by Palestinian suicide attacks. Yitzhak Rabin was assassinated on November 5, 1995 by Yigal Amir (a right-wing extremist who was against the Oslo Accords), while leaving a peace rally.

Israel withdrew from Hebron, signed the 'Wye River Memorandum' (giving greater control to the Palestinian National Authority) under Prime Minister Benjamin Netanyahu at the end of the 1990s. In 1999 Prime Minister Ehud Barak withdrew forces from Southern Lebanon and conducted negotiations with Palestinian Authority Chairman Yasser Arafat and U.S. President Bill Clinton at the 2000 'Camp David Summit'. At the summit, Barak offered a plan for a Palestinian state that included all of the Gaza Strip and over 90% of the West Bank, with Jerusalem as a shared capital. The talks failed and each side blamed the other. Likud leader Ariel Sharon controversially visited the Temple Mount and the 'Second Intifada' began. Some believe that the Second Intifada was pre-planned by Arafat as revenge for the peace talks collapse. In 2001 Sharon became prime minister in a special election and as such ended the Intifada by unilaterally withdrawing from the Gaza Strip and constructing the Israeli West Bank barrier. From 2000-2008, 1,100 Israelis had been killed, mostly by suicide bombers and the Palestinian fatalities reached 4,791 killed by Israeli security forces; 44 were killed by Israeli civilians, and 609 were killed by Palestinians.

The month-long 'Second Lebanon War' began in July 2006 after a Hezbollah artillery assault on Israel's communities on the northern border and a cross-border abduction of two Israeli soldiers. The Israeli Air Force destroyed a nuclear reactor in Syria on September 6, 2007. Another conflict ensued at the end of 2008 when a ceasefire between Hamas and Israel collapsed. The three-week '2008–09 Gaza War' ended after Israel announced a unilateral ceasefire. Hamas announced its ceasefire on the conditions of complete withdrawal and border crossing openings. The fragile ceasefire remained despite rocket launchings and Israeli retaliatory strikes continuing. On November 14, 2012 Israel responded to over a hundred rocket attacks on southern Israeli cities, by an eight day operation in Gaza. Another Gaza operation followed a Hamas escalation of rocket attacks in July 2014.

The Organization for Economic Co-operation and Development (OECD) invited Israel to join in September 2010. Israel has also signed free trade agreements with the European Union, the United States, the European Free Trade Association, Turkey, Mexico, Canada, Jordan, and Egypt. It became the first non-Latin-American country to sign a free trade agreement with the Mercosur trade bloc in 2007. From 2010-2020 regional cooperation between Israel and Arab League countries increased with: peace agreements (Jordan, Egypt); diplomatic relations with Mauritania, Bahrain, The United Arab Emirates, and Sudan; and unofficial relations with Saudi Arabia, Morocco, and Tunisia. The Arab-Israeli hostility towards Iran and its proxies changed Israel's security situation from Israel-Palestinian concerns. The post-revolutionary Islamic Republic of Iran held significant Israeli hostility since 1979 and developed into covert Iranian support of Hezbollah during the 'South Lebanon conflict' (1985–2000). The hostility essentially developed into a proxy regional conflict from 2005. Iran increased its Syrian Civil War involvement from 2011 and the conflict shifted from proxy warfare into direct confrontation by early 2018[15].

Israel faces a 'trilemma' regarding the future of the West Bank and Gaza since Israel wants:

1. to maintain control over the West Bank...it has strategic and religious significance to Israelis;
2. to retain a Jewish majority...the goal of the founding Zionist movement;
3. to remain a democracy where all citizens have full voting rights.

The continued building of Jewish settlements in the West Bank makes any separation difficult. Is Israel willing to give up on the West Bank's democracy and Jewish majority by retaining control? The book *End Game* favors partition and peace and attempts to solve the puzzle by focusing on three factors: the role of security concerns, ideology, and domestic political constraints that combine to shape Israel's strategic posture[16].

Demographically Israel is a unique country combining high fertility rates with low mortality rates and positive migration leading to a rapid rise in population. According to Taub Center projections, the country's population will reach about 12.8 million in 2040, but with distinct patterns of growth in different age groups and subpopulations. "Mortality rates are trending down, and fertility rates among Arab Israelis will fall far below 3.0 births per woman by 2040." About 80% of the annual population growth comes from very high birth rates relative to mortality rates.

The overall migration balance in Israel is positive and rising but is difficult to predict since it may occur several times in an individual's lifetime and may be influenced by historical events. Israel's population is relatively young – in the Jewish sector there are 140,000 infants versus 60,000 70-year-olds, and in the Arab Israeli sector, 42,000 versus only 5,300, respectively[17].

Israel's economic performance has been impressive. Structural reforms and huge investments in Research and Development have led to a high-tech boom. Israel is in the top ranks of the developed world's economies. Israel has several competitiveness issues that need to be addressed:

> 1. While the high-tech sector is strong, the broader Israeli economy is not accruing as many gains as it should from this strength. Too many high-tech entrepreneurs are focusing on exit strategies rather than growing their businesses into successful standalone firms. There are too few jobs being created, too few exports, and too little investment in future capacity.

> 2. Israel must improve its labor force participation rate. Low participation rates are the cause of Israel's high levels of poverty, as well as the widening social inequalities. Two groups in particular need help: the Israeli Arabs and the Ultra-Orthodox. These communities need a variety of assistance, from improved infrastructure to more market-oriented education, to vocational training and subsidized childcare. Reducing transfer payments to these groups could also provide incentives for them to join the workforce, as should reducing the number of foreign workers in the economy, as they compete with the unskilled for jobs and push down wages.

> 3. Israel's service sector has not approached its potential in terms of increased productivity growth,

rising employment, or efficient provisioning of services.

4. The public sector needs to vastly improve its transparency and efficiency. Corruption scandals are becoming all too frequent occurrences in Israel and the number and scope of regulations covering the business sector are inhibiting growth

5. Israel must continue structural reforms with labor.

6. Israel's infrastructure is merely adequate, and requires sustained long-term investments to bring its port, rail, road, and telecommunication infrastructure up to world class standards.

Making competitiveness as important as security is recommended by The Economic Strategy Institute by establishing a National Competitiveness Council linked to an Agency for Economic Development with a Technology Extension Service so low tech can be leveraged with high tech. Coordination and consensus building between labor, management, and government can be achieved with a Social Stability Pact. Create a required core national curriculum for all primary and secondary schools. Require some national service from all ethnic and religious groups in addition to military service. The Bank of Israel should manage inflation, full employment, and export growth. The tourist industry should be emphasized to provide employment to many people. The national infrastructure should be upgraded and overseen by a "czar". Reduce corporate taxes and raise short term capital gains taxes to high levels and abolish long term capital gains taxes. Impose consumption levies (Value Added Tax and a carbon tax) the proceeds of which will be used to upgrade the infrastructure. Establish a progressive flat tax for personal taxes. Work more closely with the European Union and join the European Free Trade Association (EFTA). Transfer and welfare payments need to provide incentives to get back to work by lowering direct cash payments and increasing childcare, worker retraining, adult education, and vocational programs. Infrastructure planning and spending should integrate the West Bank and Gaza[18].

References

1. *Israel's Independence Day 2019* (PDF) (Report). Israel Central Bureau of Statistics. 6 May 2019.

   Population Census 2008 (PDF) (Report). Israel Central Bureau of Statistics. 2008.

   Human Development Index and its components. United Nations Development Program 2018.

2. Charles A. Repenning & Oldrich Fejfar, *Evidence for earlier date of 'Ubeidiya, Israel, hominid site* Nature 299, 344–347 (23 September 1982).

   Encyclopedia Britannica article on Canaan

   Jonathan M Golden, *Ancient Canaan and Israel: An Introduction,* OUP, 2009 pp. 3-4.

   Abraham Malamat (1976). *A History of the Jewish People.* Harvard University Press. pp. 223–239. ISBN 978-0-674-39731-6.

   Yohanan Aharoni (15 September 2006). *The Jewish People: An Illustrated History.* A&C Black. pp. 99–. ISBN 978-0-8264-1886-9.

Declaration of Establishment of State of Israel. Israel Ministry of Foreign Affairs. 14 May 1948. Archived from the original on 17 March 2017.

Alexandrowicz, Ra'anan (24 January 2012), *The Justice of Occupation*, The New York Times.

These Are the World's Most Innovative Countries. Bloomberg.com. 24 January 2019.

Israel. Freedom in the World. Freedom House.org. 13 January, 2008.

3. *The Coastal Plain.* Israel Ministry of Tourism.

    The Living Dead Sea. Israel Ministry of Foreign Affairs. 1999. ISBN 978-0-8264-0406-0.

    Jacobs 1998, p. 284. *The extraordinary Makhtesh Ramon – the largest natural crater in the world* ...Jacobs, Daniel; Eber, Shirley; Silvani, Francesca; (Firm), Rough Guides (1998). Israel and the Palestinian Territories.

    American Friends of the Tel Aviv University, *Earthquake Experts at Tel Aviv University Turn to History for Guidance* (4 October 2007).

    Zafrir Renat, *Israel Is Due, and Ill Prepared, for Major Earthquake*, Haaretz, 15 January 2010.

    Average Weather for Tel Aviv-Yafo and Jerusalem. The Weather Channel. Archived from the original on 20 January 2013.

    Sitton, Dov (20 September 2003). *Development of Limited Water Resources – Historical and Technological Aspects.*

Israeli Ministry of Foreign Affairs. Retrieved 7 November 2007.

Grossman, Gershon; Ayalon, Ofira; Baron, Yifaat; Kauffman, Debby. *Solar energy for the production of heat Summary and recommendations of the 4th assembly of the energy forum at SNI.* Samuel Neaman Institute for Advanced Studies in Science and Technology. Archived from the original on 16 January 2013.

National Parks and Nature Reserves, Israel. Israel Ministry of Tourism. Archived from the original on 19 October 2012.

4. *Israel.* The World Factbook. Central Intelligence Agency.

    Research and development (R&D) - Gross domestic spending on R&D - OECD Data. data.oecd.org.

    These Are the World's Most Innovative Countries. Bloomberg.com.

    Moaz, Asher (2006). *Religious Education in Israel.* University of Detroit Mercy Law Review. 83 (5): 679–728.

    Israel. Academic Ranking of World Universities. 2016.

    Schwab, Klaus (2017). *The Global Competitiveness Report 2017–2018 (PDF) (Report).* World Economic Forum.

    Doing Business in Israel. World Bank Group.

    Richard Behar (11 May 2016). *Inside Israel's Secret StartupMachine.* Forbes.

    Krawitz, Avi (27 February 2007). *Intel to expand Jerusalem R&D.* The Jerusalem Post.

Berkshire Announces Acquisition. New York Times. 6 May 2006.

Israel's technology industry: Punching above its weight. The Economist. 10 November 2005.

Shteinbuk, Eduard (22 July 2011*). R&D and Innovation as a Growth Engine (PDF).* National Research University – Higher School of Economics.

Investing in Israel. New York Jewish Times. Archived from the original on 9 May 2013.

Haviv Rettig Gur (9 October 2013). *Tiny Israel a Nobel heavyweight, especially in chemistry.* The Times of Israel.

Gordon, Evelyn (24 August 2006*). Kicking the global oil habit.* The Jerusalem Post.

Stafford, Ned (21 March 2006). "Stem cell density highest in Israel". *The Scientist. Retrieved 18 October 2012.*

Futron Releases 2012 Space Competitiveness Index". Retrieved 21 December 2013.

O'Sullivan, Arieh (9 July 2012). "Israel's domestic satellite industry saved". *The Jerusalem Post.*

Space launch systems – Shavit. Deagel.

Talbot, David (2015). *Megascale Desalination.* MIT Technology Review.

Kershner, Isabel (29 May 2015). *Aided by the Sea, Israel Overcomes an Old Foe: Drought.* The New York Times. ISSN 0362-4331.

Ashkelon, Israel. water-technology.net.

Lettice, John (25 January 2008). *Giant solar plants in Negev could power Israel's future.* The Register.

Gradstein, Linda (22 October 2007). *Israel Pushes Solar Energy Technology. NPR.*

Will Israel's Electric Cars Change the World?. Time. 26 April 2011. Archived from the original on 15 April 2012.

Wainer, David; Ben-David, Calev (22 April 2010). *Israel Billionaire Tshuva Strikes Gas, Fueling Expansion in Energy, Hotels.* Bloomberg.

The World Factbook — Central Intelligence Agency. www.cia.gov.

Ketura Sun Technical Figures. Archived from the original on 9 March 2012.

Arava Power Company. Archived from original on 7 July 2011.

Roca, Marc (22 May 2012), *Arava Closes Funding For $204 Million Israeli Solar Plants.* Bloomberg.

5. *Electric Car Company Folds After Taking $850 Million From GE And Others.* Business Insider. 26 May 2013.

Roads, by Length and Area. Israel Central Bureau of Statistics. 1 September 2016.

3.09 Million Motor Vehicles in Israel in 2015. Israel Central Bureau of Statistics. 30 March 2016.

Bus Services on Scheduled Routes (PDF). Israeli Central Bureau of Statistics. 2009.

Railway Services. Israel Central Bureau of Statistics. 1 September 2016.

Burstein, Nathan (14 August 2007). *Tourist visits above pre-war level.* The Jerusalem Post.

Amir, Rebecca Stadlen (3 January 2018). *Israel sets new record with 3.6 million tourists in 2017.* Israel21.

Arava Power Company. Archived from the original on 7 July 2011.

Asian Studies: Israel as a 'Melting Pot'. National Research University Higher School of Economics. Retrieved 18 April 2012.

Mendel, Yonatan; Ranta, Ronald (2016). *From the Arab Other to the Israeli Self: Palestinian Culture in the Making of Israeli National Identity.* Routledge. pp. 137–141. ISBN 978-1-317-13171-7.

The Nobel Prize in Literature 1966. Nobel Foundation.

Israel. World Music. National Geographic Society. Archived from the original on 10 February 2012.

Davis, Barry (5 February 2007). *Israel Philharmonic Orchestra celebrates 70th anniversary.* Ministry of Foreign Affairs (from Israel21c).

History. Eurovision Song Contest. European Broadcasting Union. Retrieved 31 May 2013.

About the Red Sea Jazz Festival. Red Sea Jazz Festival. Archived from the original on 12 March 2012.

Israeli Folk Music. World Music. National Geographic Society. Archived from the original on 3 January 2012.

Freedom of the Press 2017 (PDF) (Report). Freedom House. April 2017. p. 26.

About the Museum. The Israel Museum, Jerusalem. Archived from the original on 2 March 2013.

About Yad Vashem. Yad Vashem. Archived from the original on 14 March 2012.

Ahituv, Netta (29 January 2013). *10 of Israel's best museums.* CNN.

Uzi Rebhun, Lilakh Lev Ari, *American Israelis: Migration, Transnationalism, and Diasporic Identity,* Brill, 2010 pp. 112–113.

Israel's Pork Problem. Slate. New York. 8 August 2012.

Basketball Super League Profile. Winner Basketball Super League. Retrieved 13 August 2007.

Maccabi Electra Tel Aviv – Welcome to EUROLEAGUE BASKETBALL. Archived from the original on 25 June 2014.

Chess in Schools in Israel: Progress report. FIDE. 28 May 2012.

Bekerman, Eitan (4 September 2006). *Chess masters set to blitz Rishon Letzion.* Haaretz.

Shvidler, Eli (15 December 2009). *Israeli grand master Boris Gelfand wins Chess World Cup.* Haaretz.

Israel. International Olympic Committee. Retrieved 20 March 2012.

Tel Aviv 1968. International Paralympic Committee. Archived from the original on 20 March 2012.

6. *Israel's Independence Day 2019 (PDF) (Report).* Israel Central Bureau of Statistics. 6 May 2019.

    Research and development (R&D) - Gross domestic spending on R&D - OECD Data". data.oecd.org. Retrieved 10 February 2016.

    The Land: Urban Life. Israel Ministry of Foreign Affairs. Archived from the original on 7 June 2013.

    The Law of Return. Knesset. *Archived from the original on 27 November 2005.*

    Rettig Gur, Haviv (6 April 2008). *Officials to US to bring Israelis home.* The Jerusalem Post.

    Jews, by Continent of Origin, Continent of Birth & Period of Immigration. Israel Central Bureau of Statistics. 6 September 2017.

    Goldberg, Harvey E. (2008). *From Sephardi to Mizrahi and Back Again: Changing Meanings of "Sephardi" in Its Social Environments.* Jewish Social Studies. 15 (1): 165–188. doi:10.18647/2793/JJS-2008.

    The myth of the Mizrahim. The Guardian. London. 3 April 2009.

    Shields, Jacqueline. Jewish Refugees from Arab Countries. Jewish Virtual Library.

    Israel (people). Encyclopedia.com. 2007.

Localities and Population, by Population Group, District, Sub-District and Natural Region. Israel Central Bureau of Statistics. 6 September 2017.

Population in the Localities 2019" (XLS). Israel Central Bureau of Statistics.

Settlements in the Gaza Strip". Settlement Information. Archived from the original on 26 August 2013.

Okun, Barbara S.; Khait-Marelly, Orna (2006). *Socioeconomic Status and Demographic Behavior of Adult Multiethnics: Jews in Israel* (PDF). Hebrew University of Jerusalem. Archived from the original (PDF) on 29 October 2013.

Gorenberg, Gershom (26 June 2017). *Settlements: The Real Story.* The American Prospect.

Localities and Population, by Population Group, District, Sub-District and Natural Region. Israel Central Bureau of Statistics. 6 September 2017.

7. Sanger, Andrew (2011). *The Contemporary Law of Blockade and the Gaza Freedom Flotilla. In M.N. Schmitt; Louise Arimatsu; Tim McCormack (eds.). Yearbook of International Humanitarian Law 2010. Yearbook of International Humanitarian Law. 13. p. 429.*

Israel. Academic Ranking of World Universities. 2016.

The Electoral System in Israel". The Knesset. Retrieved 8 August 2007.

Sharot, Stephen (2007). *Judaism in Israel: Public Religion, Neo-Traditionalism, Messianism, and Ethno-Religious*

Conflict. In Beckford, James A.; Demerath, Jay (eds.). The Sage Handbook of the Sociology of Religion. London and Thousand Oaks, CA: Sage Publications. pp. 671–672. ISBN 978-1-4129-1195-5.

Englard, Izhak (Winter 1987). *Law and Religion in Israel*. The American Journal of Comparative Law. 35 (1): 185–208. doi:10.2307/840166. JSTOR 840166.

The Judiciary, The Court System. Israel Ministry of Foreign Affairs. Aug 1, 2005.

Gilead Sher, *The Application of Israeli Law to the West Bank: De Facto Annexation?*, INSS Insight No. 638, 4 December 2014

Under the Guise of Security: Routing the Separation Barrier to Enable Israeli Settlement Expansion in the West Bank. Publications. B'Tselem. December 2005.

The occupied Palestinian territories: Dignity Denied. International Committee of the Red Cross. 13 December 2007.

Human Rights in Palestine and Other Occupied Arab Territories: Report of the United Nations Fact Finding Mission on the Gaza Conflict (PDF). United Nations Human Rights Council. 15 September 2009. p. 85.

Egypt: Israel must accept the land-for-peace formula. The Jerusalem Post. 15 March 2007. Retrieved 20 March 2012.

The State — Judiciary — The Court System. Israel Ministry of Foreign Affairs. 1 October 2006.

Under the Guise of Security: Routing the Separation Barrier to Enable Israeli Settlement Expansion in the West Bank. Publications. B'Tselem. December 2005.

Rudoren, Jodi; Sengupta, Somini (22 June 2015). *U.N. Report on Gaza Finds Evidence of War Crimes by Israel and by Palestinian Militants. The New York Times.*

Israel and Occupied Palestinian Territories 2016/2017. Amnesty International.

8. Ben Quinn (2017). *UK among six countries to hit 0.7% UN aid spending target.* theguardian.

   History: 1948. Israel Defense Forces. 2007. Archived from the original on 12 April 2008.

   The State: Israel Defense Forces (IDF). Israel Ministry of Foreign Affairs. 13 March 2009. Retrieved 9 August 2007.

   Shtrasler, Nehemia (16 May 2007). *"Cool law, for wrong population".* Haaretz. Retrieved 19 March 2012.

   Sherut Leumi (National Service). Nefesh B'Nefesh. Retrieved 20 March 2012.

   Arrow can fully protect against Iran. The Jerusalem Post. Retrieved 20 March 2012.

   Israeli Mirage III and Nesher Aces, By Shlomo Aloni, (Osprey 2004), p. 60

   Zorn, E.L. (8 May 2007). *Israel's Quest for Satellite Intelligence.* Central Intelligence Agency.

Proliferation of Weapons of Mass Destruction: Assessing the Risks (PDF). Office of Technology Assessment. August 1993. pp. 65, 84.

Background Information. 2005 Review Conference of the Parties to the Treaty on the Non-Proliferation of Nuclear Weapons (NPT). United Nations. 27 May 2005.

Ziv, Guy, *To Disclose or Not to Disclose: The Impact of Nuclear Ambiguity on Israeli Security*, Israel Studies Forum, Vol. 22, No. 2 (Winter 2007): 76–94

Military expenditure (% of GDP). World Development Indicators. World Bank. Retrieved 29 September 2017.

Top List TIV Tables. Stockholm International Peace Research Institute. Retrieved 21 January 2017.

9. Payne, J. P. (1996). *Book of Judges.* In Marshall, I. Howard; Millard, A. R.; Packer, J. I.; Wiseman, D. J. *(eds.).* New Bible Dictionary (3rd ed.). Leicester, England: Inter-Varsity Press. ISBN 978-0-8308-1439-8.

Thompson, Thomas L. (2000). *Early History of the Israelite People: From the Written & Archaeological Sources.* Leiden, Netherlands: Brill. ISBN 978-90-04-11943-7.

Guest, P. Deryn (2003). *Judges.* In Dunn, James D. G.; Rogerson, John William (eds.). Commentary on the Bible. Eerdmans. ISBN 978-0-8028-3711-0.

Knight, Douglas A (1995). *Deuteronomy and the Deuteronomists.* In Mays, James Luther; Petersen, David L.; Richards, Kent Harold (eds.). Old Testament Interpretation. T&T Clark. ISBN 978-0-567-29289-6.

Niditch, Susan (2008). *Judges: a commentary.* Westminster John Knox Press. ISBN 978-0-664-22096-9.

Jonathan M Golden, *Ancient Canaan and Israel: An Introduction,* OUP, 2009 pp. 3–4.

Noah Rayman (29 September 2014). *Mandatory Palestine: What It Was and Why It Matters.* TIME.

Miller, James Maxwell; Hayes, John Haralson (1986). *A History of Ancient Israel and Judah.* Westminster John Knox Press. ISBN 978-0-664-21262-9.

McNutt, Paula M. (1999). *Reconstructing the Society of Ancient Israel.* Westminster John Knox. ISBN 978-0-664-22265-9.

Miller, Robert D. (2012) [First published 2005]. *Chieftains of the Highland Clans.* ISBN 978-1-62032-208-6.

Levine, Robert A. (7 November 2000). *See Israel as a Jewish Nation-State, More or Less Democratic.* The New York Times.

Kramer, Gudrun (2008). *A History of Palestine: From the Ottoman Conquest to the Founding of the State of Israel.* Princeton University Press. ISBN 978-0-691-11897-0.

Palestine – Ottoman rule. www.britannica.com. Encyclopedia Britannica.

Mandate for Palestine, Encyclopedia Judaica, Vol. 11, p. 862, Keter Publishing House, Jerusalem, 1972

Eisen, Yosef (2004). *Miraculous journey: a complete history of the Jewish people from creation to the present.* Targum Press. ISBN 978-1-56871-323-6.

Jewish and Non-Jewish Population of Palestine-Israel (1517–2004). Jewish Virtual Library. Retrieved 29 March 2010.

Immigration to Israel. Jewish Virtual Library. Retrieved 29 March 2012.

10. *Palestine in the Hellenistic Age - My Jewish Learning.* myjewishlearning.com

    Lemche, Niels Peter (1998). *The Israelites in History and Tradition.* Westminster John Knox Press. p. 35. ISBN 978-0-664-22727-2.

    Grabbe, Lester L. (2004). *A History of the Jews and Judaism in the Second Temple Period: Yehud – A History of the Persian Province of Judah v. 1.* T & T Clark. ISBN 978-0-567-08998-4.

    Oppenheimer, A'haron and Oppenheimer, Nili. *Between Rome and Babylon: Studies in Jewish Leadership and Society.* Mohr Siebeck, 2005, p. 2.

11. Gil, Moshe (1997). *A History of Palestine, 634–1099.* Cambridge University Press. ISBN 978-0-521-59984-9.

    Benzion Dinur (1974). *From Bar Kochba's Revolt to the Turkish Conquest.* In David Ben-Gurion (ed.). *The Jews in their Land.* Aldus Books.

    Samson ben Abraham of Sens, *Jewish Encyclopedia.*

    Kramer, Gudrun (2008). *A History of Palestine: From the Ottoman Conquest to the Founding of the State of Israel.* Princeton University Press. ISBN 978-0-691-11897-0.

Dan Bahat (1976). *Twenty centuries of Jewish life in the Holy Land: the forgotten generations.* Israel Economist. p. 48.

Fannie Fern Andrews (1976). *The Holy Land under mandate.* Hyperion Press. p. 145. ISBN 978-0-88355-304-6.

Joel Rappel, History of Eretz Israel from Prehistory up to 1882 (1980), vol. 2, p. 531.

Palestine – Ottoman rule. www.britannica.com. Encyclopedia Britannica.

Mandate for Palestine, Encyclopedia Judaica, Vol. 11, p. 862, Keter Publishing House, Jerusalem, 1972

12. *https://www.jewishvirtuallibrary.org/the-diaspora*

    Mandate for Palestine, Encyclopedia Judaica, Vol. 11, p. 862, Keter Publishing House, Jerusalem, 1972

    Geoffrey Wigoder, G.G. (ed.). *Return to Zion.* The New Encyclopedia of Judaism.

    Gilbert 2005, p. 2. *Jews sought a new homeland here after their expulsions from Spain (1492) ...*

    Jewish and Non-Jewish Population of Palestine-Israel (1517-2004). Jewish Virtual Library.

    Immigration to Israel. Aliyah During World War II and its Aftermath. Jewish Virtual Library.

    Kornberg 1993 How did Theodor Herzl, an assimilated German nationalist in the 1880s, suddenly in the 1890s become the founder of Zionism?

Macintyre, Donald (26 May 2005). *The birth of modern Israel: A scrap of paper that changed history.* The Independent.

Schechtman, Joseph B. (2007). *Jewish Legion.* Encyclopaedia Judaica. 11. Detroit: Macmillan Reference. p. 304.

League of Nations: *The Mandate for Palestine, July 24, 1922.* Modern History Sourcebook. 24 July 1922.

Walter Laqueur (2009). *A History of Zionism: From the French Revolution to the Establishment of the State of Israel.* Knopf Doubleday Publishing Group. ISBN 978-0-307-53085-1.

13. *Killing Centers: An Overview.* Holocaust Encyclopedia. United States Holocaust Memorial Museum.

Documenting Numbers of Victims of the Holocaust and Nazi Persecution. United States Holocaust Memorial and Museum. 4 February 2019.

Crowe, David M. (2008). *The Holocaust: Roots, History, and Aftermath.* Boulder, CO: Westview Press. ISBN 978-0-8133-4325-9.

Friedlander, Henry (1994). *Step by Step: The Expansion of Murder, 1939-1941.* German

Studies Review 17 (3): 495-507. doi:10.2307/1431896. JSTOR 1431896.

Friedländer, Saul (1997). *Nazi Germany and Jews: The Years of Persecution 1933 to 1939.*

New York: Harper Collins. ISBN 0-06-019042-6.

Niewyk, Donald L.; Nicosia, Francis R. (2000). *The Columbia Guide to the Holocaust.*

New York: Columbia University Press. ISBN 0-231-11200-9.

Proctor, Robert (1988). *Racial Hygiene: Medicine Under the Nazis.* Cambridge, MA:

Harvard University Press. ISBN 0-674-74578-7.

Burleigh, Michael (2001). *The Third Reich: A New History.* New York: Hill and Wang. ISBN 0-8090-9326-X.

Black, Jeremy (2016). *The Holocaust: History and Memory.* Bloomington, IN: Indiana University Press. ISBN 978-0-253-02214-1.

Persecution of Homosexuals in the Third Reich. Holocaust Encyclopedia. United States Holocaust Memorial Museum. *Archived* from the original on 12 June 2018.

Non-Jewish Resistance. Holocaust Encyclopedia. United States Holocaust Memorial Museum. *Archived* from the original on 20 November 2011.

Reparations and Restitutions (PDF). Shoah Resource Center. Yad Vashem. *Archived (PDF)* from the original on 16 May 2017.

Payment Program of the Foundation EVZ. Bundesarchiv. Archived from the original on 5 October 2018.

Staff (29 May 2013). *Holocaust Reparations: Germany to Pay 772 Million Euros to Survivors.* Spiegel Online International. Archived from the original on 13 December 2014.

Davies, Lizzie (17 February 2009). *France responsible for sending Jews to concentration camps, says court.* The Guardian. *Archived* from the original on 10 October 2017.

14. *Resolution 181 (II). Future government of Palestine.* United Nations. 29 November 1947. Retrieved 21 March 2017.

Declaration of Establishment of State of Israel. Israel Ministry of Foreign Affairs. 14 May 1948. Archived from the original on 17 March 2017.

The Population of Palestine Prior to 1948. MidEastWeb. Retrieved 19 March 2012.

Motti Golani (2013). *Palestine Between Politics and Terror, 1945–1947.* UPNE. p. 130. ISBN 978-1-61168-388-2.

Cohen, Michael J (2014). *Britain's Moment in Palestine: Retrospect and Perspectives, 1917–1948 (First ed.).* Abingdon and New York: Routledge. ISBN 978-0-415-72985-7.

Clarke, Thurston. *By Blood and Fire*, G.P. Puttnam's Sons, New York, 1981

A/364. Special Committee on Palestine. United Nations. 3 September 1947. Archived from the original on 10 June 2012.

Gelber, Yoav (2006). *Palestine 1948.* Brighton: Sussex Academic Press. p. 17. ISBN 978-1-902210-67-4.

Tal, David (2003). *War in Palestine, 1948: Israeli and Arab Strategy and Diplomacy.* Routledge. p. 471. ISBN 978-0-7146-5275-7.

Jacobs, Frank (7 August 2012). *The Elephant in the Map Room". Borderlines.* The New York Times.

Karsh, Efraim (2002). *The Arab-Israeli conflict: The Palestine War 1948.* Osprey Publishing. p. 50. ISBN 978-1-84176-372-9.

David Tal (2004). *War in Palestine, 1948: Israeli and Arab Strategy and Diplomacy.* Routledge. p. 469. ISBN 978-1-135-77513-1.

Karsh, Efraim (2002). *The Arab-Israeli conflict: The Palestine War 1948.* Osprey Publishing. ISBN 978-1-84176-372-9.

Morris, Benny (2004). *The Birth of the Palestinian Refugee Problem Revisited.* Cambridge University Press. p. 602. ISBN 978-0-521-00967-6.

15. *The Kibbutz & Moshav: History & Overview.* Jewish Virtual Library.

    Anita Shapira (1992). *Land and Power.* Stanford University Press. pp. 416, 419.

    Segev, Tom. 1949: The First Israelis. *The First Million.* Trans. Arlen N. Weinstein. New York: The Free Press, 1986. pp. 105-107

    Shulewitz, Malka Hillel (2001). *The Forgotten Millions: The Modern Jewish Exodus from Arab Lands.* Continuum. ISBN 978-0-8264-4764-7.

    Population, by Religion. Israel Central Bureau of Statistics. 2016.

Kameel B. Nasr (1996). *Arab and Israeli Terrorism: The Causes and Effects of Political Violence, 1936–1993.* McFarland. pp. 40–. ISBN 978-0-7864-3105-2. ^ Gilbert 2005, p. 58

Adolf Eichmann. Jewish Virtual Library.

John Quigley, *The Six-Day War and Israeli Self-Defense: Questioning the Legal Basis for Preventive War,* Cambridge University Press, 2013, p. 32.

Andrews, Edmund; Kifner, John (27 January 2008). *George Habash, Palestinian Terrorism Tactician, Dies at 82".* The New York Times. Retrieved 29 March 2012.

1973: Arab states attack Israeli forces. On This Day. BBC News. 6 October 1973.

Basic Law: Jerusalem, Capital of Israel. Knesset.

Golan Heights profile. BBC News. 27 November 2015.

Friedberg, Rachel M. (November 2001). *The Impact of Mass Migration on the Israeli Labor Market (PDF).* The Quarterly Journal of Economics. 116 (4): 1373–1408.

The Oslo Accords, 1993. U.S. Department of State.

The Wye River Memorandum. U.S. Department of State. 23 October 1998.

Ain, Stewart (20 December 2000). *PA: Intifada Was Planned.* The Jewish Week.

Harel, Amos; Issacharoff, Avi (1 October 2010). *Years of rage.* Haaretz.

Harel, Amos (13 July 2006). *Hezbollah kills 8 soldiers, kidnaps two in offensive on northern border.* Haaretz.

Ravid, Barak (18 January 2009). *IDF begins Gaza troop withdrawal, hours after ending 3-week offensive.* Haaretz.

Lappin, Yaakov; Lazaroff, Tovah (12 November 2012). *Gaza groups pound Israel with over 100 rockets.* The Jerusalem Post.

Israel and Hamas Trade Attacks as Tension Rises. The New York Times. 8 July 2014.

Israel's Free Trade Area Agreements, IL: Tamas, archived from the original on 3 October 2011.

16. www.brookings.edu › book › end-game

17. taubcenter.org.il › pr-eng-population-projections-israel-.

18. https://www.econstrat.org/research/country-and-area-studies/354-israel-2020-a-strategic-vision-for-economic-development

Made in United States
North Haven, CT
06 June 2022